BIOFUELS IN INDIA
A NEW REVOLUTION

Dr. Kumari Swarnim

MANGLAM
MANGLAM PUBLICATIONS
DELHI-110053 (INDIA)

Published by :
MANGLAM PUBLICATIONS
L-21/1, Street No. 5, Shivaji Marg, Near Kali Mandir,
J.P. Nagar, Kartar Nagar, West Ghonda, Delhi-110053
Phone : 011-22945677, (M.) 9968367559, 9868572512
E-mail: manglam.books2007@rediffmail.com
manglam.publications2007@rediffmail.com

Biofuels in India: A New Revolution

First Edition 2010

ISBN 978-93-80013-48-0

PRINTED IN INDIA

Published by Ravindra Kumar Yadav for Manglam Publications, Delhi-110053 Laser Typesetting at M.K. Graphics, Delhi- 53, Printed at Sachin Printers, Moujpur Delhi- 53

Preface

The world is on the verge of an unprecedented increase in the production and use of biofuels for transport. The combination of rising oil prices, issues of security, climate instability and pollution, deepening poverty in rural and agricultural areas, and a host of improved technologies, is propelling governments to enact powerful incentives for the use of these fuels, which is in turn sparking investment.

This book presents an overview of the current and future markets for biodiesels in India. It also provides feedstock analysis, market drivers, restraints, and future strategies for the industry. In this research, Frost & Sullivan's expert analysts thoroughly examine the biodiesel and bioethanol markets.

Held back by the lack of large-scale availability of feedstock, the Indian biodiesel market trails its global counterparts by a long way. It is likely to take a while for biodiesel to be established as an effective biofuel, since Jatropha plantations in the country are still in the initial stages of development. Three to four years and many plantations later, the country may have the feedstock necessary for the large-scale production of Jatropha oil for use in biodiesel. The absence of a clear Government policy on Jatropha oil production has inhibited several biofuel manufacturers from entering this market. Hence, Indian manufacturers are considering importing palm oil to produce biodiesel.

The better-developed Indian bioethanol market is also grappling with similar availability issues, as ethanol is primarily manufactured from molasses - a by-product of sugar. Since sugarcane production is cyclical, the availability and cost of production of bioethanol will vary depending on sugarcane crop yields. India's ethanol-blending program could not be implemented during 2003-2004 due to a low sugarcane output and the second phase of this program was announced in September 2006 only after a recovery in sugarcane production. Overall, the Government and industry have to show greater initiatives toward the Jatropha program to help biodiesel manufacturers save costs. Meanwhile, in the bioethanol sector, further research is necessary to aid in the development of alternate feedstock and improvement in production efficiency.

A strong economy, rising incomes, and a vibrant market have given a huge boost to the transport sector in India, which is the fastest growing energy-consuming sector in the country. This sector's energy demand is expected to grow by 6 to 8 percent per annum during the 11th five-year plan period (2007-2012). With more than 80 percent of passengers and 60 percent of freight being transported by road, it is obvious that the dependence of personal modes of transport, such cars and two-wheelers, has increased drastically. "The automotive vehicle population is growing by 12 to 15 percent per annum and this will, in turn, impact the transport sector's energy demand," says the analyst of this research. "Diesel and gasoline (petrol) contribute to 98 percent of the energy consumed in the transport sector."

India's crude oil and petroleum products supplies are largely import-dependant. With oil import expenditure increasing by more than six times in the last 25 years due to escalation in global demand and prices, biofuels are likely to be pressed into service. This alternate form of fuel will be critical in reducing the dependence on fossil fuels, achieving greater energy security, and reducing noxious emissions. "The Government is currently implementing an ethanol-blending program, while it is also considering initiatives in the form of mandates for biodiesel," notes the analyst. "Due to these mandates, the rising population, and the growing energy demand from the transport sector, biofuels will be assured of a significant market in India."

—Author

Contents

1

Introduction

A critical need exists to investigate various renewable and solar energy technologies and examine the energy and environmental issues associated with these various technologies. The various renewable energy technologies will not be able to replace all current 102 quads (quad = 1015 BTU) of U.S. energy consumption (USCB 2007). A gross estimate of land and water resources is needed, as these resources will be required to implement the various renewable energy technologies.

The world, and the United States in particular, face serious energy shortages and associated high energy prices during the coming decades. Oil, natural gas, coal, and nuclear power provide more than 88% of world energy needs; the other 12% is provided by various renewable energy sources. Oil, natural gas, coal, and nuclear provide more than 93% of U.S. energy needs; the other 9% consists of various renewable and non-renewable energy sources. The U.S., with slightly more than 45% of the world's population, accounts for nearly 25% of the world's energy consumption. On average, each American uses nearly 8,000 L of oil equivalents per year for all purposes, including transportation, industry, heating and cooling.

The United States now imports more than 63% of its oil at an annual cost of approximately $200 billion (USCB 2007). Projections are that within 20 years the U.S. will be importing more than 90% of its oil. The United States has consumed more than 90% of its proved oil reserves (Pimentel et al. 2004a). Because the U.S. population is growing nearly twice as fast as that of China per capita, and is adding 3.3 million to the population each year, energy resources are becoming scarce (PRB 2006). These shortages are now contributing to greater interest in renewable energy resources.

Diverse renewable energy sources currently provide 6.8% of U.S. needs and about 12% of world needs. In addition to energy conservation, the development and use of renewable energy is expected to increase as fossil fuel supplies decline and become highly expensive. Eight different renewable technologies are projected to provide the U.S. with most of its energy in the future: hydropower, biomass, wind power, solar thermal, photovoltaics, passive energy systems, geothermal, and biogas. In this chapter, I assess the potential of these 8 renewable energy technologies, including their environmental benefits and risks, and their energetic and economic costs.

HYDROELECTRIC POWER

Hydropower contributes significantly to world energy, providing 6% of the supply. In the United States, hydroelectric plants produce approximately 3% or 3.4 quads of total U.S. energy (340 billion kWh) (1 kWh = 860 kilocalories [kcal] = 3,440 BT = 3.6 megajoules), or 11% of the nation's electricity, each year at a cost of $0.02 per kWh. Development and rehabilitation of existing dams in the United States could produce an additional 5 quads per year. Hydroelectric plants, however, require considerable land for their water storage reservoirs. An average of 75,000 hectares (ha) of reservoir land area and 14 trillion L of water are required per 1 billion kWh per year produced. Based on regional estimates of US land use and average annual energy generation, reservoirs currently cover approximately 26 million ha of the total 917 million ha of land area in the United States (Pimentel 2001).

To develop the remaining best candidate sites, assuming land requirements similar to those in past developments, an additional 7 million ha of land would be required for water storage. Despite the benefits of hydroelectric power, the plants cause major environmental problems. The impounded water frequently covers valuable, agriculturally productive, alluvial bottomland. Sediments build up behind the dams, reducing their effectiveness and creating another major environmental problem. Further, dams alter the existing plants, animals, and microbes in the ecosystem (Nilsson and Berggren 2000). Fish species may significantly decline in river systems because of these numerous ecological changes.

BIOMASS ENERGY

Most biomass is burned for cooking and heating, however, it can also be converted into electricity and liquid fuel. Under sustainable forest conditions in both temperate and tropical ecosystems, approximately 3 dry metric tons (t/ha) per year of woody biomass can be harvested sustainably (Birdsey 1992, Repetto 1992, Trainer 1995, Ferguson 2003). Although this amount of woody biomass has a gross energy yield of 13.5 million kcal/ha, it requires an energy expense of approximately 33 L of diesel fuel per ha, plus the embodied energy for cutting and collecting wood for transport to an electric power plant. Thus, the energy input per output ratio for a woody biomass system is calculated to be 1:22.

The cost of producing 1 kWh of electricity from woody biomass is about $0.06, which is competitive with other electricity production systems that average $0.07 in the U.S. (USCB 2007). Approximately 3 kWh of thermal energy is expended to produce 1 kWh of electricity, an energy input/output ratio of 1:7. Per capita consumption of woody biomass for heat in the United States amounts to 625 kilograms (kg) per year. In developing nations, use of diverse biomass resources (wood, crop residues, and dung) average about 630 kg per capita (Kitani 1999). Developing countries use only about 500 L of oil equivalents of fossil energy per capita compared with nearly 8,000 L of oil equivalents of fossil energy used per capita in the United States.

Woody biomass could supply the United States with about 5 quads (1.5 × 1012 kWh thermal) of its total gross energy supply by the year 2050, provided there was adequate forest land available. A city of 100,000 people using the biomass from a sustainable forest (3 t/ha per year) for electricity would require approximately 200,000 ha of forest area, based on an average electrical demand of slightly more than 1 billion kWh (electrical energy [e]) (860 kcal = 1 kWh). Environmental impacts of burning biomass are less harmful than those associated with coal, but more harmful than those associated with natural gas (Pimentel 2001). Biomass combustion releases more than 200 different chemical pollutants, including 14 carcinogens and 4 co-carcinogens, into the atmosphere (Burning Issues 2003). Globally, but especially in developing nations where people cook with fuelwood over open fires, approximately 4 billion people suffer from continuous exposure to smoke (Kids for Change 2006). In the United States, wood smoke kills 30,000 people each year (EPA, 2002). However, the pollutants from electric plants that use wood and other biomass can be controlled.

WIND POWER

For many centuries, wind power has provided energy to pump water and to run mills and other machines. Today, turbines with a capacity of at least 500 kW produce most of the commercially wind-generated electricity. Operating at an ideal location, one of these turbines running at 30% efficiency can yield an energy output of 1.3 million kWh (e) per year (AWEA 2000a). An initial investment of approximately $500,000 for a 500kW capacity turbine operating at 30% efficiency, will yield an input/output ratio of 1:4 over 30 years of operation. During the 30-year life of the system, the annual operating costs amount to about $50,000. The estimated cost of electricity generated is $0.07 per kWh (e). Some report costs ranging from $0.03 to $0.05 per kWh (Sawin 2004). These values are probably located in favorable wind sites.

In the United States, 2502 megawatts (MW) of installed wind generators produce about 6.6 billion kWh of electrical energy per year (Chambers 2000). The American Wind Energy Association (AWEA 2000b) estimates that the United States could support a capacity of 30,000MW by the year 2010, producing 75 billion kWh (e) per year at a capacity of 30%, or approximately 2% of the annual US electrical consumption. If all economically feasible land sites are developed, the full potential of wind power is estimated to be about 675 billion kWh (e) (AWEA 2000b). Offshore sites could provide an additional 102 billion kWh (e) (Gaudiosi 1996), making the total estimated potential of wind power 777 billion kWh (e), or 23% of current electrical use.

Widespread development of wind power is limited by the availability of sites with sufficient wind (at least 20 kilometers per hour [km/h]) and the number of wind machines that the site can accommodate. An average area for one 50kW turbine is 1.3 ha to allow sufficient spacing to produce maximum power. Based on this figure, approximately 9,500 ha of land are needed to supply 1 billion kWh per year. Because the turbines themselves only occupy approximately 2% of the area, most of the land can be used for vegetables, nursery stock, and cattle (Natural Resources Canada 2002). However, it may be impractical to produce corn or other grains because of the heavy equipment used in this type of farming. An investigation of the environmental impacts of wind energy production reveals a few hazards. Locating the wind turbines in or near the flyways of migrating birds and wildlife refuges may result in birds flying into the supporting towers and rotating blades.

For this reason, it is suggested that wind farms be located at least 300 meters (m) from nature reserves to reduce their risk to birds. The estimated 13,000 wind turbines installed in the United States kill an estimated 2,600 birds per year (Sinclair 2003). Choosing a proper site and improving repellant technology with strobe lights or paint patterns might further reduce the number of birds killed. Bat fatalities are another serious concern. It is projected that by 2020 annual bat fatalities caused by wind turbines will range from 33,000 to 62,000 individuals annually (Kunz et al. 2007). Most bat fatalities are from species that migrate long distances and are tree roosting. Eastern U.S wind turbines installed along forested ridgetops have the highest rate of bat kills, ranging from 15.3 to 41.1 bats per MW of installed capacity per year (Kunz et al. 2007). Monitoring for bat and bird fatalities and research for the reduction of these should be included in all wind energy planning.

The rotating magnets in the turbine electrical generator produce a low level of electromagnetic interference that can affect television and radio signals within 2–3 km of large installations (Sagrillo 2006). Fortunately, with the widespread use of cable networks or line-of-sight microwave satellite transmission, both television and radio are unaffected by this interference. The noise caused by rotating blades is another unavoidable side effect of wind turbine operation. Beyond 2.1 km, however, the largest turbines are inaudible even downwind. At a distance of 400 m, the noise level is estimated to be about 60 decibel, corresponding roughly to the noise level of a home air-conditioning unit.

SOLAR THERMAL CONVERSION SYSTEMS

Solar thermal energy systems collect the sun's radiant energy and convert it into heat. This heat can be used directly for household and industrial purposes or produce steam to drive turbines that produce electricity. The complexity of these systems ranges from solar ponds to electricity-generating parabolic troughs. In the following analyses, I convert thermal energy into electricity to facilitate comparison to the other solar energy technologies.

SOLAR PONDS

Solar ponds are used to capture radiation and store the energy at temperatures of nearly 100 degrees Celsius (°C). Constructed ponds can be made into solar ponds by creating a layered salt concentration gradient. The layers prevent natural convection, trapping the heat collected from solar radiation in the bottom layer of brine. The hot brine from the bottom of the pond is piped out to use for heat, for generating electricity, or both.

For successful operation of a solar pond, the salt concentration gradient and the water level must be maintained. A solar pond covering 4,000 ha loses approximately 3 billion L of water per year (750,000 L/ha per year) under arid conditions (Tabor and Doran 1990). Recently, solar ponds in Israel have been closed because of such difficulties. To counteract the water loss and the upward diffusion of salt in the ponds, the dilute salt water at the surface of the ponds has to be replaced with fresh water and salt added to the lower layer (Solar Pond 2007). The efficiency of solar ponds in converting solar radiation into heat is estimated to be approximately 1:4, assuming a 30-year life for the solar pond. A 100 ha (1 km2) solar pond can produce electricity at a rate of approximately $0.30 per kWh (Australian Government 2007).

Some hazards are associated with solar ponds, but most can be avoided with careful management. It is essential to use plastic liners to make the ponds leakproof and prevent contamination of the adjacent soil and groundwater with salt.

PARABOLIC TROUGHS

Another solar thermal technology that concentrates solar radiation for large-scale energy production is the parabolic trough. A parabolic trough, shaped like the bottom half of a large drainpipe, reflects sunlight to a central receiver tube that runs above it. Pressurized water and other fluids are heated in the pipe and used to generate steam that drives turbogenerators for electricity production or provides heat energy for industry.

Parabolic troughs that have entered the commercial market have the potential for efficient electricity production because they can achieve high turbine inlet temperature. Assuming peak efficiency and favorable sunlight conditions, the land requirements for the central receiver technology are approximately 1,100 ha per1 billion kWh per year. The energy input:output ratio is calculated to be 1:5. Solar thermal receivers are estimated to produce electricity at approximately $0.07–$0.09 per kWh (DOE/EREN 2001).

The potential environmental impacts of solar thermal receivers include the accidental or emergency release of toxic chemicals used in the heat transfer system. Water availability can also be a problem in arid regions.

PHOTOVOLTAIC SYSTEMS

Photovoltaic cells have the potential to provide a significant portion of future U.S. and world electrical energy (Energy Economics 2007). Photovoltaic cells produce electricity when sunlight excites electrons in the cells. The most promising photovoltaic cells in terms of cost, mass production, and relatively high efficiency are those manufactured using silicon. Because the size of the unit is flexible and adaptable, photovoltaic cells can be used in homes, industries, and utilities. However, photovoltaic cells need improvements to make them economically competitive before their use can become widespread. Test cells have reached efficiencies of about 25% (American Energy 2007), but the durability of photovoltaic cells must be lengthened and current production costs reduced several times to make their use economically feasible.

Production of electricity from photovoltaic cells currently costs about $0.25 per kWh (DOE 2000). Using mass-produced photovoltaic cells with about 18% efficiency, 1 billion kWh per year of electricity could be produced on approximately 2,800 ha of land, and this is sufficient electrical energy to supply 100,000 people. Locating the photovoltaic cells on the roofs of homes, industries, and other buildings would reduce the need for additional land by an estimated 20% and reduce transmission costs. However, because storage systems such as batteries cannot store energy for extended periods, photovoltaics require conventional backup systems.

The energy input for making the structural materials of a photovoltaic system capable of delivering 1 billion kWh during a life of 30 years is calculated to be approximately 143 million kWh. Thus, the energy input per output ratio for the modules is about 1:7.

The major environmental problem associated with photovoltaic systems is the use of toxic chemicals, such as cadmium sulfide and gallium arsenide, in their manufacture. Because these chemicals are highly toxic and persist in the environment for centuries, disposal and recycling of the materials in inoperative cells could become a major problem.

GEOTHERMAL SYSTEMS

Geothermal energy uses natural heat present in Earth's interior. Examples are geysers and hot springs, like those at Yellowstone National Park in the United States. Geothermal energy sources are divided into three categories: hydrothermal, geopressured-geothermal, and hot dry rock. The hydrothermal system is the simplest and most commonly used for electricity generation. The boiling liquid underground is produced using wells, high internal pressure drives, or pumps. In the United States, nearly 3,000 MW of installed electric generation comes from hydrothermal resources, and this is projected to increase by 4,500MW.

Most of the geothermal sites for electrical generation are located in California, Nevada, and Utah. Electrical generation costs for geothermal plants in the West range from $0.06 to $0.30/kWh (Gawlik and Kutscher 2000), suggesting that this technology offers potential to produce electricity economically. The US Department of Energy and the Energy Information Administration (DOE/EIA

2001) project that geothermal electric generation may grow three- to fourfold during the next 20–40 years. However, other investigations are not as optimistic and, in fact, suggest that geothermal energy systems are not renewable because the sources tend to decline over 40–100 years (Bradley 1997, Youngquist 1997, Cassedy 2000). Existing drilling opportunities for geothermal resources are limited to a few sites in the United States and world (Youngquist 1997).

Potential environmental problems of geothermal energy include water shortages, air pollution, waste effluent disposal, subsidence, and noise. The wastes produced in the sludge include toxic metals such as arsenic, boron, lead, mercury, radon, and vanadium.Water shortages are an important limitation in some regions. Geothermal systems produce hydrogen sulfide, a potential air pollutant; however, this could be processed and removed for use in industry. Overall, these environmental costs of geothermal energy appear to be minimal relative to those of fossil fuel systems.

BIOGAS

Wet biomass materials can be converted effectively into usable energy using anaerobic microbes. In the United States, livestock dung is normally gravity fed or intermittently pumped through a plug-flow digester, which is a long, lined, insulated pit in the earth. Bacteria break down volatile solids in the manure and convert them into methane gas (65%) and CO2 (35%) (Pimentel 2001). A flexible liner stretches over the pit and collects the biogas, inflating like a balloon. The biogas may be used to heat the digester, to heat farm buildings, or to produce electricity. A large facility capable of processing the dung from 500 cows costs nearly $300,000 (EPA 2000). The Environmental Protection Agency (EPA 2000) estimates that more than 2000 digesters could be economically installed in the United States.

The amount of biogas produced is determined by the temperature of the system, the microbes present, the volatile solids content of the feedstock, and the retention time. A plug-flow digester with an average manure retention time of about 16 days under winter conditions (17.4°C) produced 452,000 kcal/day and used 262,000 kcal/day to heat the digester to 35°C (Jewell et al.

1980). Using the same digester during summer conditions (25°C) but reducing the retention time to 10.4 days, the yield in biogas was 524,000 kcal/day, and it used 157,000 kcal/day for heating the digester (Jewell et al. 1980). The energy input per output ratios for these winter and summer conditions for the digester were 1:1.7 and 1:3.3, respectively. The energy output of biogas digesters is similar today (Hartman et al. 2000). In developing countries such as India, biogas digesters typically treat the dung from 15 to 30 cattle from a single family or a small village. The resulting energy produced for cooking saves forests and preserves the nutrients in the dung. The capital cost for an Indian biogas unit ranges from $500 to $900 (Kishore 1993). The price value of a kWh biogas in India is about $0.06 (Dutta et al. 1997). The total cost of producing about 10 million kcal of biogas is estimated to be $321, assuming the cost of labor to be $7/h; hence, the biogas has a value of $356. Manure processed for biogas has fewer odors and retains its fertilizer value (Pimentel 2001).

ETHANOL AND ENERGY INPUTS

In the fermentation/distillation process, the corn is finely ground and approximately 15 L of water are added per 2.69 kg of ground corn. After fermentation, to obtain a liter of 95% pure ethanol from the 8% ethanol and 92% water mixture, the 1 L of ethanol must be extracted from the approximately 13 L of the ethanol/water mixture. To be mixed with gasoline, the 95% ethanol must be further processed and more water removed, requiring additional fossil energy inputs to achieve 99.5% pure ethanol. Thus, a total of about 12 L of wastewater must be removed per liter of ethanol produced, and this relatively large amount of sewage effluent has to be disposed of at an energy, economic, and environmental cost.

To produce a liter of 99.5% ethanol uses 43% more fossil energy than the energy produced as ethanol and costs 44c/ per L ($1.66 per gallon or $2.76 per gallon including the subsidy). The corn feedstock requires more than 33% of the total energy input. In this analysis the total cost, including the energy inputs for the fermentation/distillation process and the apportioned energy costs of the stainless steel tanks and other industrial materials, is $436.92

per 1,000 L of ethanol produced.

The largest energy inputs in corn-ethanol production are for producing the corn feedstock, plus the steam energy, and electricity used in the fermentation/distillation process. The total energy input to produce a liter of ethanol is approximately 7,570 kcal. However, a liter of ethanol has an energy value of only 5,130 kcal. Based on a net energy loss of 2,440 kcal of ethanol produced, 43% more fossil energy is expended than is produced as ethanol.

GRASSLANDS AND CELULOSIC ETHANOL

Tilman's research (Tillman et al. 2006) has merit in the explanation of field experiments with various combinations of species of natural vegetation, and the productivity of diverse experimental systems. The outstanding, 30-year effort by the Land Institute in Kansas (Jackson 1980) to develop multi-species perennial ecosystems that deliver high productivity for long periods has been *de facto* endorsed by Tillman et al., albeit without acknowledgement.

However, there are concerns about two items. First, the statement by Tillman et al. that crop residues, like corn stover, can be harvested and utilized as a fuel source. This would be a disaster for agricultural ecosystems. Without the protection of crop residues, soil loss may increase as much as 100-fold (Fryrear and Bilbro 1994). Already the U.S. crop system is losing soil 10 times faster than sustainability (NAS 2003). Soil formation rates are extremely slow or less than 1 t/ha/yr (NAS 2003, Troeh et al. 2004). Increased erosion will facilitate soil-C oxidation and contribute to the greenhouse problem (Lal 2003). Tillman et al. assume about 1,032 L of ethanol can be produced through the conversion of the 4 t/ha/yr of grasses harvested. However, Pimentel and Patzek (2007) reported a negative 50% return in switchgrass conversion. Based on the optimistic data of Tillman et al., and converting all 235 million ha of U.S. grassland into ethanol, only 12% of U.S. petroleum would be provided (USDA 2004, USCB 2004–2005).

In addition, to achieve the production of this much ethanol would mean displacing about 100 million cattle, 7 million sheep, and 4 million horses now grazing on 324 million ha of U.S. grassland and rangeland (USDA 2004, Mitchell 2000). Already overgrazing is a problem on U.S. grasslands and a similar problem exists worldwide (Brown 2001). Thus, the assessment of the quantity of ethanol that can be produced on U.S. and world grasslands by Tillman et al. appears to be unduly optimistic.

METHANOL AND VEGETABLE OILS

Methanol can be produced from a gasifier-pyrolysis reactor using biomass as a feedstock (Hos and Groenveld 1987, Jenkins 1999). The yield from 1 ton of dry wood is about 370 L of methanol (Ellington et al. 1993, Osburn and Osburn 2001). For a plant with economies of scale to operate efficiently, more than 1.5 million ha of sustainable forest would be required to supply it (Pimentel 2001). Biomass is generally not available in such enormous quantities, even from extensive forests, at acceptable prices. Most methanol today is produced from natural gas. Processed vegetable oils from soybean, sunflower, rapeseed, and other oil plants can be used as fuel in diesel engines. Unfortunately, producing vegetable oils for use in diesel engines is costly in terms of economics and energy (Pimentel and Patzek 2005). A slight net return on energy from soybean oil is possible, if the soybeans are grown without commercial nitrogen fertilizer. The soybean under favorable conditions will produce its own nitrogen. Even assuming a slight net energy return with soy, the total United States would have to be planted to soybeans just to provide soy oil for U.S. trucks!

TRANSITION TO RENEWABLE ENERGY

Despite its environmental and economic benefits, the transition to large-scale use of renewable energy presents several difficulties. Renewable energy technologies, all of which require land for collection and production, will compete with agriculture, forestry, and urbanization for land in the United States and world. The United States is at maximum use of its prime cropland for food production per capita today, but the world has less than half the cropland per capita that it needs for a diverse diet (0.5 ha) and adequate supply of essential nutrients (USDA 2004). In fact, more than 3.7 billion people are already malnourished in the world (UN/SCN 2004, Bagla 2003). With the world and US populations expected to double in the next 58 and 70 years, respectively, all the available cropland and forestland will be required to provide vital food and forest products (PRB 2006).

As the growing U.S. and world populations demand increased electricity and liquid fuels, constraints like land availability and high investment costs will restrict the potential development of renewable energy technologies. Energy use based on current growth is projected to increase from the current U.S. consumption of 102 quads to approximately 145 quads by 2050. Land availability is also a problem, with the US population adding about 3.3 million people each year (USCB 2007). Each person added requires about 0.4 ha (1 acre) of land for urbanization and highways and about 0.5 ha of cropland (Vesterby and Krupa 2001). Renewable energy systems require more labor than fossil energy systems. For example, wood-fired steam plants require several times more workers than coal-fired plants (Giampietro et al. 1998).

An additional complication in the transition to renewable energies is the relationship between the location of ideal production sites and large population centers. Ideal locations for renewable energy technologies are often remote, such as deserts of the American Southwest or wind farms located kilometers offshore. Although these sites provide the most efficient generation of energy, delivering this energy to consumers presents a logistical problem. For instance, networks of distribution cables must be installed, costing about $179,000 per km 115-kV lines (DOE/EIA 2002). A percentage of the power delivered is lost as a function of electrical resistance in the distribution cable. There are complex alternating current electrical networks in North America, and 3 of these are tied together by DC lines (Nordel 2001). Based on these networks, it is estimated that electricity can be transmitted up to 1500 km.

A sixfold increase in installed technologies would provide the United States with approximately 46 quads (thermal) of energy, less than half of current US consumption. This level of energy production would require about 159 million ha of land (17% of US land area). This percentage is an estimate, and could increase or decrease depending on how the technologies evolve and energy conservation is encouraged.

Worldwide, approximately 473 quads of all types of energy are used by the population of more than 6.5 billion people/ Using available renewable energy technologies, an estimated 200 quads of renewable energy could be produced worldwide on about 20% of the world land area. A self-sustaining renewable energy system producing 200 quads of energy per year for about 2 billion people (Ferguson 2001) would provide each person with about 5,000 L of oil equivalents per year, approximately half of America's current consumption per year, but an increase for most people of the world (Pimentel et al. 1999).

The first priority of the US energy program should be for individuals, communities, and industries to conserve fossil fuel resources and reduce consumption. Other developed countries have proved that high productivity and a high standard of living can be achieved with the use of half the energy expenditure of the United States (Pimentel et al. 1999). In the United States, fossil energy subsidies of approximately $40 billion per year should be withdrawn and the savings invested in renewable energy research and education to encourage the development and implementation of renewable technologies. If the United States became a leader in the development of renewable energy technologies, then it would likely capture the world market for this industry (Shute 2001).

The current subsidies for ethanol production total $6 billion per year (Koplow 2006). This means that the subsidies per gallon of ethanol are 60 times greater than the subsidies per gallon of gasoline!

CONCLUSION

This assessment of renewable energy technologies confirms that these techniques have the potential to provide the nation with alternatives to meet nearly half of future U.S. energy needs. To develop this potential, the United States would have to commit to the development and implementation of non-fossil fuel technologies and energy conservation. People in the U.S. would have to reduce their current energy consumption by more than 50% and this is entirely possible. Eventually we will be forced to reduce energy consumption. The implementation of renewable energy technologies now would reduce many of the current environmental problems associated with fossil fuel production and use.

The United States' immediate priority should be to speed the transition from the reliance on nonrenewable fossil energy resources to reliance on renewable energy technologies. Various combinations of renewable technologies should be developed consistent with the characteristics of the different geographic regions in the United States. A combination of the renewable technologies should provide the United States with an estimated 46 quads of renewable energy by 2050. These technologies should be able to provide this much energy without interfering with required food and forest production.

If the United States does not commit itself to the transition from fossil to renewable energy during the next decade or two, the economy and national security will suffer. It is of critical importance that U.S. residents work together to conserve energy, land, water, and biological resources. To ensure a reasonable standard of living in the future, there must be a fair balance between human population density and use of energy, land, water, and biological resources.

2

Potential of Earth to Deliver the Biomass for Fuel

INTRODUCTION

The purpose of this work is to: 1. Show that the current and proposed "cellulosic" ethanol (a "second generation" agrofuel) refineries are inefficient, low energy-density concentrators of solar light. 2. Prove that even if these refineries were marvels of efficiency, they still would be able to make but a dent in our runaway consumption of transportation fuels, because the Earth simply has little or no biomass to spare in the long run. The fundamental energy unit I use in this work is 1 exajoule (EJ) or 1018 joules

A little over four joules heats one teaspoon of water by 1 degree Celsius. One statistical American develops average continuous power of almost exactly 100W (Patzek, 2007). One exajoule in the digested food feeds amply 300 million people1 for one year. The actual food available for consumption in the US is ca. 2 EJ yr″1, and the entire food system uses <″20 EJ yr″1 (Patzek, 2007). Currently, Americans are using about 105 EJ yr″1 (340 GJ (yr-person)″1), or 105 times more primary energy than needed as food. The EU countries use 80 EJ yr″1 of primary energy or 55% less energy per capita than US. Current consumption of all transportation fuels in the US is about 33 EJ yr″1.

A barely visible fraction of this energy comes from corn ethanol. According to current government plans, the amount of ethanol produced in the US will reach 35 billion gallons in 2017, but it is difficult to imagine that a 30 billion gallon per year increase will come from corn ethanol. Before peaking in 2006, the world production of conventional petroleum grew exponentially at 6.6% per year between 1880 and 1970. The Hubbert curves are symmetrical (Patzek, 2007) and predict world production of conventional petroleum to decline exponentially at a similar rate within a decade from now, or so. This decline can be arrested for a while by heroic measures (infill drilling, horizontal wells, enhanced oil recovery methods, etc.), but the longer it is arrested the more precipitous it will become. If the current per capita use of petroleum in the US is escalated with the expected growth of US population, the US will have to intercept the *entire* estimated production of conventional petroleum3 in the world by 2042. In this scenario, the projected *increment* of US petroleum consumption between today and 2042 is equivalent to 270 billion gallons of ethanol per year.

BACKGROUND

Humans *are* an integral part of a single system made of all life and all parts of the Earth's near-surface. Thus, as President Vaclav Havel said on July 4, 1994: "Our destiny is not dependent merely on what we do to ourselves but also on what we do for [the Earth] as a whole. If we endanger her, she will dispense with us in the interest of a higher value - life itself." So how to proceed? It appears that humanity's survival is subject to these five constraints:

- **Constraint 1:** An almost exponential rate of growth of human population.
- **Constraint 2:** Too much use of Earth resources; in particular, fossil fuels; and even more specifically, liquid transportation fuels.
- **Constraint 3:** The Earth that is too small to feed in perpetuity 7 billion people and counting, 1 billion cows, and – now – 1 billion cars.
- **Constraint 4:** The ossified political structures in which more is better, and more of the same is also safer.
- **Constraint 5:** A global climate change.

Unfortunately, these five constraints prevent existence of a *stable continuous* solution to human life in the near-future.

Alternatively, we may choose from the following two *discontinuous* solutions:

- **Solution 1:** Extinguish ourselves and much of the living Earth, or
- **Solution 2:** Fundamentally and abruptly change, while slowly decreasing our numbers.

Problems with Change The last time humanity ran mostly on living plant carbon was approximately in 1760. There was 1 billion of us, and we certainly knew how to feed ourselves due to the latest Agricultural Revolution that started in Europe a century earlier (Osborne, 1970). Our food supply problems then had to do with political madness, inaptitude, and greed – just as they do today (Davis, 2002). Today, however, there is almost 7 times more of us. We can still feed ourselves, but with huge inputs of fossil carbon in addition to fresh plant carbon, minerals, and soil. These inputs also mine fossil water and pollute surface water, aquifers, the oceans, and the atmosphere.

By extrapolating human population growth between 1650 and 1920 to 2007, one estimates 2.2 billion people who today could live mostly on plant carbon, but use some coal, oil, and natural gas. Therefore it is reasonable to say that today 4.5 billion people owe their existence to the Haber-Bosch ammonia process and the fossil fuel-driven, fundamentally unstable "Green Revolution," as well as to vaccines and antibiotics. Agrofuels are a direct outgrowth of the "Green Revolution," which may be viewed, as a short-lived but violent disturbance of terrestrial ecosystems on the Earth.

Since most people have cooked or ridden in a vehicle, many feel empowered to talk about *energy* as though they were experts. It turns out, however, that issues of energy supply, use, environmental impacts, and – especially – of *free* energy are too complicated for the *adlib* homilies we hear every day in the media. Professor Vardaraja Raman, a well-known physicist and humanist, said it best: "A major problem confronting society is the lack of knowledge among the public as to what science is, what constitutes scientific thinking and analysis, and what science's criteria are for determining the correctness of statements about the phonomenological world."

It is a misconception that Constraint 2 can be removed with fresh plant carbon, while forgetting the scale of Constraint 1 and ignoring Constraint 3. Constraint 4 helps us to maintain that more biomass converted to liquid fuels means more of the same lifestyles, and a stable continuation of the current socioeconomic systems – Constraint 3 be damned. Constraint 5 plays the role of a wild card. Its unknown negative impacts may dwarf everything else I have mentioned in this work.

This leads me to the first major conclusion of this work: Business as usual will lead to a complete and practically immediate crash of the technically advanced societies and, perhaps, all humanity. This outcome will not be much different from a collapse of an overgrown colony of bacteria on a petri dish when its sugar food runs out and waste products build up. Today, the human "petri dish" is Earth's surface, and "food" is the living matter and water we consume and the ancient plant products and minerals (oil, natural gas, coal, etc.) we mine and burn.

The Earth operates in *endless* cycles and modern humans race along *short* line segments. At each turn of her cycles, the Earth renews herself, but humans are about to wake up inside a huge toxic waste dump with nowhere to go.

PLAN OF ATTACK

As you are beginning to suspect, it is not sufficient to limit ourselves just to discussing liquid transportation fuels and their future biological sources. These transportation fuels intrude upon every other aspect of life on the Earth: Availability of clean water to drink and clean air to breathe, healthy soil and healthy food supply, destruction of biodiversity and essential planetary services in the tropics, acceleration of global climate change, and so on.

As with many important policy-making decision processes, I start from the end, here the cellulosic ethanol refineries. This is where most public money, attention, and hope are. I show that these refineries are inefficient compared with the existing petroleum- and corn-based refineries, and are difficult to scale up. Then I return to the beginning and show that even *if* the cellulosic biomass refineries were marvels of efficiency, they still could *not* maintain our current lifestyles by a long stretch, simply because the Earth will not give us the extra biomass needed to keep on existing as we do. For a while we might continue to *rob* this biomass from the poor tropics, but the results are already disastrous for all humanity.

EFFICIENCY OF CELLULOSIC ETHANOL REFINERIES

I start from a "reverse-engineering" calculation of energy efficiency of cellulosic ethanol production in an *existing* Iogen pilot plant, Ottawa, Canada. I then discuss the inflated energy efficiency claims of five out-of-six recipients of $385 millions of DOE grants to develop cellulosic ethanol refineries.

IOGEN OTTAWA FACILITY

Wheat, oat, and barley straw are first pretreated with sulfuric acid and steam. Iogen's patented enzyme then breaks the cellulose and hemicelluloses down into six- and five-carbon sugars, which are later fermented and distilled into ethanol. Normal yeast does not ferment the 5-carbon sugars, so genetically modified, delicate and patented yeast strains are used. Iogen's plant has capacity of 1 million gallons of ethanol per year.

Since wheat is the largest grain crop in Canada, I use its straw as a reference (the other two straws are similar). On a dry mass basis (dmb), wheat straw has 33% of cellulose, 23% of hemicelluloses, and 17% total lignin.8. Other sources report 38%, 29%, and 15% dmb, respectively, see (Lee et al., 2007) for a data compliation. These differences are not surprising, given experimental uncertainties and variable biomass composition. To calculate ethanol yield, I use the more optimistic, second set of data.

The calculated ethanol yield, 0.18 kg EtOH (kg straw dmb)"1, is somewhat less than a recently reported maximum ethanol yield of 0.24 kg/kg (Saha et al., 2005) achieved in 500mL vessels, starting from 48.6% of cellulose. Simultaneous saccharification and fermentation yielded 0.17 kg/kg. Because enzymatic decomposition of cellulose and hemicelluloses is inefficient, the resulting dilute broth requires 2.4 times more energy to distill than the average 15 MJL"1 in an average ethanol refinery (Patzek, 2004; Patzek, 2006a).

One may argue that Iogen's Ottawa facility is for demonstration purposes only and that the saccharification and fermentation batches were not regularly scheduled. In this case, an alternative calculation yields the same result: At about 0.2 to 0.25 kg of straw/L, the mash is barely pumpable.With Badger's yield of 0.18 kg/kg of EtOH, the highest ethanol yield is 3.5 – 4.4% of ethanol in water.

The higher heating value (HHV) of ethanol is 29.6MJ kg"1 (Patzek, 2004). The HHV of wheat straw is 18.1 MJ kg"1 (Schmidt et al., 1993) and that of lignin 21.2MJ kg"1 (Domalski et al., 1987). With these inputs the first-law (energy) efficiency of Iogen's facility is ç = 0.28 × 29.6 1 × 18.1 + 0.18 × 2.4 × 15/0.787 " 0.15 × 21.2 H" 20% (2.1) where the density of ethanol is 0.787 kg L"1 and the entire HHV of lignin was used to offset distillation fuel (another optimistic assumption for the wet separated lignin). This calculation disregards the energy costs of high-pressure steam treatments of the straw at 120 or 140°C, and the separated solids at 190°C, sulfuric acid and sodium hydroxide production, etc. Also, the complex enzyme production processes must use plenty of energy.

This analysis leads to the second conclusion: The Iogen plant in Ottawa, Canada, has operated well below name plate capacity for three years. Iogen should retain their trade secrets, but in exchange for the significant subsidies from the US and Canadian taxpayers they should tell us what the annual production of alcohols *was*, how much straw was used, and what the fossil fuel and electricity inputs were. The ethanol yield coefficient in kg of ethanol per kg straw dmb is key to public assessments of the new technology. Similar remarks pertain to the Novozymes projects heavily subsidized by the Danes. Until an existing pilot plant provides real, independently verified data on yield coefficients, mash ethanol concentrations, etc., all proposed cellulosic ethanol refinery designs are *speculation*.

PROPOSED CELLULOSIC ETHANOL REFINERIES

Now I present at *face value* the stated energy efficiencies9 of the six proposed10 cellulosic ethanol plants awarded 385 million USD by the US Department of Energy. For calibration, after 87 years of development and optimization, the actual energy efficiency of Sasol's Fischer-Tropsch coal-to-liquid fuels plants is about 42% (Steynberg and Nel, 2004). The average energy efficiency of the highly optimized corn ethanol refineries is 37% (not counting grain coproducts as fuels). An average petroleum refinery is about 88% energy-efficient.11 For details, see (Patzek, 2006a,b,c) The DOE/USDA report by Perlack et al. (2005) has led to the claims by an influential venture capitalist, Mr. Vinod Khosla (2006), of being able to produce 130 billion gallons of ethanol from 1.4 billion tons of biomass (dmb), apparently at a 52% thermodynamic efficiency.

To see how very different the new fossil-energy-free world will be, let's compare power from Iogen's plant with that from an oil well in the US. Ever more *power* is what we must have to continue our current way of life. Iogen's plant delivers the power of 7 barrels of oil per day (68 kW). Average power of petroleum wells in the largely oil-depleted US was 10 bbl (well-day)" in 200612 (98 kW). Therefore, an average US petroleum well delivers more power than a city-block size Iogen facility in Ottawa and its area of straw collection, probably 50 km in radius, which at this time is saturated with fossil fuels outright and their products (ammonia fertilizers, field chemicals, roads, etc.). The petroleum well also uses little input power; unfortunately, soon petroleum will not be a transportation option. Such is the difference between solar energy stocks (depletable fossil fuels) and flows (daily photosynthesis). One can calculate that an average agricultural worker in the US uses 800kW of fossil energy inputs and outputs 3,000kW. An average oil & gas worker in California uses 2,800kW of fossil energy inputs and outputs 14,500kW. Due to fossil energy and machines these two workers are supermen, each capable of doing the work of 8,000 and 28,000 ordinary humans, respectively. These two fellows are about to become human again, and we need to get used to this idea.

Where will the Agrofuel Biomass Come from?

Collectively, the EU and the US have spent billions of dollars to be able to construct the inefficient behemoth factories, which in the distant future might ingest megatonnes or gigatonnes of apparently free biomass "trash" and spit out priceless liquid transportation fuels. It is therefore prudent to ask the following question: Call out using the new paragraph and gray background. The answer to this question is immediate and unequivocal: Nowhere, close to nothing, and for a very short time indeed. On the average, our planet has **zero** excess biomass at her disposal.

Useful Terminology

Several different ecosystem13 productivities, i.e., measures of biomass accumulation per unit area and unit time have been used in the ecological literature, e.g., (Reichle et al., 1975; Randerson et al., 2001) and many others. Usually this biomass is expressed as grams of carbon (C) per square meter and per year, or as grams of water-free biomass (dmb) per square meter and year. The conversion factor between these two estimates is the carbon mass fraction in the fundamental building blocks of biomass, CH*x*O*y* , where *x* and *y* are real numbers, e.g., 1.6 and 0.6, that express the overall mass ratios of hydrogen and oxygen to carbon. The following definitions are common in ecology:

1. *Gross* Primary Productivity, GPP = mass of CO2 fixed by plants as glucose.
2. Ecosystem respiration, *Re* = mass of CO2 released by metabolic activity of autotrophs, *Ra,* and heterotrophs (consumers and decomposers), *Rh*: *Re* = *Ra* + *Rh* (2.2) where decomposers are defined as worms, bacteria, fungi, etc. Plants respire about 1/2 of the carbon available from photosynthesis after photorespiration, with the remainder available for growth, propagation, and litter production, see (Ryan, 1991). Heterotrophs respire most, 82–95%, of the biomass left after plant respiration (Randerson et al., 2001).
3. *Net* Primary Productivity, NPP = GPP "*Ra.*
4. Net *Ecosystem* Productivity NEP = GPP " *Re* " Non " *R* sinks and flows (2.3) The older NEP definitions would usually neglect the non respiratory losses, e.g., (Reichle et al., 1975). All ecological definitions of NEP I have seen, lump incorrectly mass flows and mass sources and sinks, calling them "fluxes," see, e.g., (Randerson et al., 2001; Lugo and Brown, 1986).

Plant Biomass Production

The reason for the Earth recycling all of her material parts can be explained. The Earth is powered by the sun's radiation that crosses the outer boundary of her atmosphere and reaches her surface. The Earth can export into outer space long-wave infrared radiation.15 But, because of her size, the Earth holds on to all mass of all chemical elements, except perhaps for hydrogen. By maintaining an oxygen-rich atmosphere, *life* has managed to prevent the airborne hydrogen from escaping Earth's gravity by reacting it back to water (and destroying ozone). If all mass must stay on the Earth, all her households must recycle everything; otherwise internal chemical waste would build up and gradually kill them. Mother Nature does not usually do toxic waste landfills and spills.

In a mature ecosystem, one species' waste must be another species' food and no *net* waste is ever created. The little imperfections in the Earth's *surface* recycling programs have resulted in the burial of a remarkably *tiny* fraction of plant carbon in swamps, lakes, and shallow coastal waters. Very rarely the violent anoxic events would kill most of life in the oceanic waters and cause faster carbon burial. Over the last 460,000,000 years (and going back all the way to 2,500,000,000 years ago), the Earth has gathered and transformed *some* of the buried ancient plant mass into the fossil fuels we love and loath so much. The proper mass balance of carbon fluxes in terrestrial ecosystems, confirms the compelling, thermodynamic argument that sustainability of any ecosystem requires *all* mass to be conserved on the average. The larger the spatial scale of an ecosystem and the longer the time-averaging scale are, the stricter adherence to this rule must be. Such are the laws of nature.

Physics, chemistry and biology say clearly that there can be *no* sustained net mass *output* from any ecosystem for more than a few years. A young forest in a temperate climate grows fast in a clear-cut area, and transfers nutrients from soil to the young trees. The young trees grow very fast (there is a positive NPP), but the amount of mass accumulated in the forest is small. When a tree burns or dies some or most of its nutrients go back to the soil. When this tree is logged and hauled away, almost no nutrients are returned. After logging young trees a couple of times the forest soil becomes depleted, while the populations of insects and pathogens are well-established, and the forest productivity rapidly declines (Patzek and Pimentel, 2006).When the forest is allowed to grow long enough, its net ecosystem productivity becomes zero on the average.

Therefore, in order to export biomass (mostly water, but also carbon, oxygen, hydrogen and a plethora of nutrients) an ecosystem must import equivalent quantities of the chemical elements it lost, or decline irreversibly. Carbon comes from the atmospheric CO2 and water flows in as rain, rivers and irrigation from mined aquifers and lakes. The other nutrients, however, must be rapidly *produced* from ancient plant matter transformed into methane, coal, petroleum, phosphates, etc., as well as from earth minerals (muriate of potash, dolomites, etc.), – all *irreversibly* mined by humans. Therefore, to the extent that humans are no longer integrated with the ecosystems in which they live, they are doomed to extinction by exhausting all planetary stocks of minerals, soil and clean water. The question is not *if*, but *how fast*? It seems that with the exponentially accelerating *mining* of global ecosystems for biomass, the time scale of our extinction is shrinking with each crop harvest. Compare this statement with the feverish proclamations of sustainable biomass and agrofuel production that flood us from the confused media outlets, peer-reviewed journals, and politicians.

Is There any Other Proof of NEP = *0*?

I just gave you an abstract proof of no trash production in Earth's Kingdom, except for its dirty human slums. Are there any other, more direct proofs, perhaps based on measurements? It turns out that there are two approaches that complement each other and lead to the same conclusions. The first approach is based on a top-down view of the Earth from a satellite and a mapping of the reflected infrared spectra into biomass growth. I will summarize this proof here. The second approach involves a direct counting of all crops, grass, and trees, and translating the weighed or otherwise measured biomass into net primary productivity of ecosystems. Both approaches yield very similar results.

Satellite Sensor-Based Estimates

Global ecosystem productivity can be estimated by combining remote sensing with a carbon cycle analysis. The US National Aeronautics and Space Administration (NASA) Earth Observing System (EOS) currently "produces a regular global estimate of gross primary productivity (GPP) and annual net primary productivity (NPP) of the entire terrestrial earth surface at 1-km spatial resolution, 150 million cells, each having GPP and NPP computed individually" (Running et al., 2000). The MOD17A2/ A3 User's Guide (Heinsch et al., 2003) provides a description of the Gross and Net Primary Productivity estimation algorithms (MOD17A2/A3) designed for the MODIS sensor.

NPP in the US The overall median values of net primary productivity may be converted to the higher heating value (HHV) of NPP in the US. In 2003, thus estimated net annual biomass production in the US was 5.3 Gt and its HHV was 90 EJ. One must be careful, however, because the underlying distributions of ecosystem productivity are different for each ecosystem and highly asymmetric. Therefore, lumping them together and using just one median value can lead to a substantial systematic error. For example, the lumped value of US NPP of 90 EJ, underestimates the overall

To limit this error, one can perform a more detailed calculation based on the 16 classes of land cover (Hurtt et al., 2001). The MODIS-derived median NPPs are reported for most of these classes. Since the spatial set of land-cover classes cannot be easily mapped onto the administrative set of USDA classes of cropland, woodland, pastureland/ rangeland, and forests, Hurtt et al. (2001) provide an approximate linear mapping between these two sets, in the form of a 16 × 4 matrix of coefficients between 0 and 1. I have lumped the land-cover classes somewhat differently (to be closer to USDA's classes). The Cropland + Mosaic class here comprises the USDA's cropland, woodland, and some of the pasture classes. The Remote Vegetation class comprises some of the USDA's rangeland and pastureland classes. The USDA forest class is somewhat larger than here, as some of the smaller patches of forest, such as parks, etc., are in the Mosaic class. Thus calculated 2003 US NPP is 118 EJ yr"1, 74EJ yr"1 of above-ground (AG) plant construction and 44 EJ yr"1 in root construction. In addition 12/ 74 = 17% of AG vegetation is in remote areas, not counting the remote forested areas. Note that my use of land-cover classes and their typical root-to-shoot ratios yields an overall result (118 EJ yr"1) which is very similar to that derived by the Numerical Terradynamic Simulation Group (113 EJ yr"1).

Therefore, the DOE/USDA proposal to produce 130 billion gallons of ethanol from 1400 million tonnes of biomass (Perlack et al., 2005) each year – and year-after-year –, would consume 32% of the remaining above-ground NPP in the US, if one assumes a 52% energy-efficiency of the conversion. At the current 26% overall efficiency of the corn-ethanol cycle (Patzek, 2006a), roughly 64% of all AG NPP in the US would have to be consumed to achieve this goal with zero harvest losses. To use more than half of all accessible above-ground plant growth in all forests, rangeland, pastureland and agriculture in the US to produce agrofuels would be a continental-scale ecologic and economic disaster of biblical proportions. Conclusions

I have shown that the Earth simply cannot produce the vast quantities of biomass we want to use to prolong our unsustainable lifestyles, while slowly committing suicide as a global human civilization.

In passing, I have noted that the "cellulosic biomass" refineries are very inefficient, currently impossible to scale, and incapable of ever catching up with the runaway need to feed one billion gasoline- and diesel-powered cars and trucks.

3

Rewards and Risks of Ethanol Production

Ethanol production doubled in a very short period of time in the U.S. due to a combination of natural disasters, political tensions, and much more demand globally from petroleum. Responses to this expansion will span many sectors of society and the economy. As the Midwest gears up to rapidly add new ethanol manufacturing plants, the existing regional economy must accommodate the changes.

There are issues for decision makers regarding existing agricultural activities, transportation and storage, regional economic impacts, the likelihood of growth in particular areas and decline in others, and the longer term economic, social, and environmental sustainability. Many of these issues will have to be considered and dealt with in a simultaneous fashion in a relatively short period of time.

This chapter investigates sets of structural, industrial, and regional consequences associated with ethanol plant development in the Midwest, primarily, and in the nation, secondarily. The first section untangles the rhetoric of local and regional economic impact claims about biofuels.

The second section describes the economic gains and offsets that may accrue to farmers, livestock feeding, and other agri-businesses as production of ethanol and byproducts increase. The last section discusses the near and longer term growth prospects for rural areas in the Midwest and the nation as they relate to biofuels production.

INTRODUCTION

The economic, social, political, and environmental impacts of modern ethanol production in the U.S. are highly regionalized. Current ethanol production and most new ethanol plant development in the United States are concentrated in the Corn Belt states of Iowa, Illinois, Indiana, Minnesota, and Nebraska. Those states alone produced nearly 62 percent of the nation's corn in 2006. Not surprisingly, those same states account for about two-thirds of actual or planned ethanol production capacity. Ethanol production and plant development took on an added urgency in the fall of 2005 after hurricanes Katrina and Rita crippled domestic oil production capacity in the Gulf of Mexico. Those events, coupled with heightened uncertainty about both near-term and long-term oil supplies in light of other international issues, fueled massive amounts of rhetorical, political, and financial resources in support of biofuels production and energy independence.

The growth in U.S. ethanol production has been dramatic: In 2005, 1.6 billion bushels of corn were converted to ethanol, about 12.1 percent of the total corn supply. By the end of 2007 it is estimated that 3.2 billion bushels will be used for that purpose, about a quarter of the nation's corn supply, and an increase of just over 100 percent in only two years (USDA 2007). That much corn will make enough ethanol to account for 3.9 percent of the nation's total demand for motor gasoline that year (EIA 2007). Expansion in ethanol production from corn through the rest of this decade is expected to top out at from 4.0 billion bushels by 2010 (USDA 2007) to 4.3 billion bushels (FAPRI 2007), though some analysts can envision sets of policy and market considerations that might push production higher (Tokgoz et al. 2007).

Responses to this expansion in ethanol production will span many sectors of society and the economy. Already, the expansion in production capacity has driven up corn prices sharply from recent historical levels, which in turn has driven up the number of acres planted in corn: 2007 corn acres nationally are 19 percent higher than 2006. But given a generally fixed supply of arable farmland, there are consequences to this expansion: soybean plantings declined by 15 percent and cotton by 28 percent (USDA June 2007). Over the past two decades, national farm commodity production has been a relatively stable, slowly-adjusting mix of crops and livestock with very distinct regional advantages and production concentrations. The rapid rise in ethanol production from corn, however, likens to dropping a large rock in a calm pond – there are ripples extending in all directions that affect crop production, animal production, food production, and, ultimately, the well-being of households. As the Midwest gears up to rapidly add new ethanol manufacturing plants, the existing regional economy must accommodate the changes. There are issues for decision makers regarding existing agricultural activities, transportation and storage, regional economic impacts, the likelihood of growth in particular areas and decline in others, and the longer term economic, social, and environmental sustainability. Many of these issues will have to be considered and dealt with in a simultaneous fashion in a relatively short period of time.

This chapter investigates sets of structural, industrial, and regional consequences associated with ethanol plant development in the Midwest, primarily, and in the nation, secondarily. The first section untangles the rhetoric of local and regional economic impact claims about biofuels. The second section describes the economic gains and offsets that may accrue to farmers, livestock feeding, and other agri-businesses as production of ethanol and byproducts increase. The last section discusses the near and longer term growth prospects for rural areas in the Midwest and the nation as they relate to biofuels production.

MEASURING AND MISMEASURING BIOFUELS ECONOMIC IMPACTS

It is important to sort out the rhetoric of claimed economic benefits to be expected from biofuels development in the Midwest and the nation because there are tremendous amounts of public money at stake. In the very early stages of this modern boom in ethanol plant construction, politicians, farm commodity groups, and economic developers hailed the emerging industry as the right and proper evolution of modern agricultural production capacities coupled inexorably with technological breakthroughs and long overdue changes in the nation's energy policies. Amidst this enthusiasm, biofuels trade associations and some agricultural commodity groups reported in various venues that scores of thousands of jobs have been created across the Corn Belt and the nation. Some politicians and government agency representatives parroted those reports uncritically; Midwestern state governments began to specifically and energetically apply government agency services in support of the boom, along with offering lucrative tax credits and incentives to promote even faster growth; land-grant universities promoted their vital scientific contributions in this coming energy revolution; cities and counties scrambled to be the site of a modern ethanol factory, to be on the plus side of economic trends for a change given the historical deterioration of rural Midwestern economies and communities; and some leaders in Midwestern states began to envision a social and economic resurgence in rural areas.

Profound expectations like the aforementioned demand careful scrutiny, especially when massive amounts of national, state, and local government subsidy are at stake. The place to begin is with the measurement of net economic gain attributable to this run-up in ethanol production in the U.S. and the identification of who benefits. Those aggressively promoting private and public investment in more biofuels processing capacities range from farm commodity groups, farm state politicians, some environmental organizations, automobile manufacturers, to both liberal and conservative political orientations.

There are wide ranges of economic activity attributed to biofuels production. The nation's production of ethanol creates jobs at the ethanol plants, boosts the demand for critical mechanical, technical, and service inputs, and helps to improve the prices received by input commodity providers, namely corn producers. Beyond that, few of the conclusions about the economic impacts of biofuels production appear to be based on rigorous, enterprise or industry level research, however (Swenson 2006). Much is of a very rudimentary level using broad assumptions about ethanol industry activity and applying, uncritically and often inappropriately, national economic impact ratios to deduce the size of economic activity attributable to ethanol production. The estimates either at the local level or at the national level are quite diverse and often incredible.

As examples, at the national level, an Urbanchuck (2005) report for the Renewable Fuels Association used US Bureau of Economic Analysis factors to conclude that 114,844 jobs in the national economy depended indirectly on the operation of all ethanol plants and the purchases that are made by workers (and this did not include ethanol plant employment). Earlier in the decade, when the industry was even smaller, Novack (2002) of the Federal Reserve Bank of Kansas City was more upbeat about the job total and reported in a widely read periodical that ". . . the [ethanol] industry added nearly 200,000 jobs to the U.S. economy." This is a curious claim given that the U.S. Department of Commerce's industrial census for that same year (2002) indicated the ethyl-alcohol industry had just 2,200 jobs. How the author got from 2,200 jobs to 200,000 is not revealed, but the writer went on to predict that "an additional 214,000 jobs [would] be created through the economy over the next decade." Last, as just one example of comments made by many farm state politicians, former South Dakota U.S. Senator Thomas Daschle concluded in a national and widely reprinted publication that the production of 3.1 billion gallons of ethanol in the U.S. created 200,000 jobs (Daschle 2006).

These three examples are emblematic of the rhetoric underscoring ethanol production expansion and public policy development in the U.S. The first was made by a consultancy with long-standing ties with the Renewable Fuels Association, a trade group that aggressively promotes corn ethanol policies and serves as the primary information source for information on renewable fuels opportunities and capacities in the U.S. The second claim came from a writer from the nation's respected public banking regulatory and financial research sector. In this case the Kansas City Federal Reserve Bank also has a specialization in rural development economic studies and affairs; hence, an assumption of rigor and credibility. The third job claim came from a respected and long-time political leader and strong advocate for alternative energy development. Given the implied authority of these three sources it is important to investigate the source of their numerical enthusiasms. A good example for understanding the basis for the robust, yet quite misleading, job claims can be found in recent work sponsored by the Iowa Renewable Fuels Association.

DECONSTRUCTING ETHANOL JOB IMPACT CLAIMS IN THE MIDWEST

An Urbanchuck (2007) report for the Iowa Renewable Fuels Association (IRFA) concluded that Iowa's ethanol industry had created 46,938 jobs and contributed $7.315 billion in state domestic product. Research at Iowa State University (Swenson 2007b) concluded, in contrast, that the state's 28 ethanol producers in processing 600 million bushels of corn into approximately 1.65 billion gallons of ethanol created from 4,100 to 4,700 net new jobs in the Iowa economy through 2005. The public university statistics are a tenth of those produced by the trade group. The following exercise explains most of the differences.

First, from the original number of 46,938 jobs are subtracted the 19,733 jobs linked to capital development and construction. There are several good reasons for doing this: Those are not net new permanent jobs – the jobs were all ready in the larger regional economy as there is a generally fixed rate of capital formation in the U.S. linked to the availability of investment resources and the overall pace and pattern of capital growth; according to U.S. Bureau of Economic Analysis statistics, the overall national rate of investment in the chemical manufacturing industry where ethanol is located is actually less than the average for all manufacturing for the 2000–2005 period; there is a finite number of plants that can and will be built given this state's current and likely future supply of corn and the rate of national absorption of ethanol; and the capital development that those construction workers are contributing to serves significantly as substitutes for energy-related and other forms of industrial development in Iowa, the greater region, and in the nation. Eliminating the existing and spatially temporary construction jobs leaves us 27,205 jobs to further parse.

Next, a full two-thirds of the purported non-construction ethanol impact jobs were already in the economy whether there was or there was not an ethanol industry. The IRFA study used a set of final demand multipliers to estimate the remaining ethanol job and product impacts (BEA 1997). Final demand means that either the industry is producing for final consumption by households and institutional users within the region or it is producing for consumption by entities external to the region of production. The fundamental assumption in the use of a final demand multiplier and its interpretation, however, is that expansion in ethanol production creates, concomitantly and at fixed rates, expansions in all inter-industrial relations that industry has with all of its inputs suppliers. So the use of a final demand multiplier for a particular industry, like the organic chemical industry where ethanol production is located assumes that as that industry expands production, there are fixed-ratio expansions in all industries that provide its intermediate inputs.

There is a fundamental flaw here because there is no real change in the overall demand for corn in the short run, just a shift in corn deliveries destined for local processing instead of for export. As a consequence, the application of a final demand multiplier to the corn sector is completely spurious. Those jobs already existed and would have existed had there not been an expansion in Iowa ethanol facilities. The ethanol plant did not create the corn production jobs or all of the corn industry's up-stream supply linkages. To claim them as ostensibly having been created by the emerging ethanol industry is misleading. To reiterate: ethanol production is not creating more farmers.

So from the 27,205 total jobs attributed to Iowa's ethanol industry operations in the RFA report we must next subtract the 18,398 jobs linked to its existing corn production sector. That leaves 8,807 jobs to investigate. Several other items of critical inputs into production into this industry that are listed in the IRFA study after the already discounted corn values must be scrutinized. First, and importantly, the Iowa ethanol industry requires a large amount of natural gas, electricity, and water. The job gains attributable in that study to these three industries combined for 2,591 of the remaining 8,807 potential ethanol economic impact jobs. Those utility suppliers, however, are massive, declining cost industries in which the average costs of delivering their respective commodities up to capacity decline sharply. An industry that is an extremely heavy, and therefore comparatively easy to supply, user of a particular commodity is delivered that commodity at a substantially reduced price due to strong distributional efficiencies. Large users of utilities do not stimulate average job multiplier effects – they stimulate much lower, marginal effects and as a consequence are charged rates that are significantly lower than those charged to smaller users. This is a fundamental flaw in fixed-ratio impact analysis employed by the authors of the study and one of the reasons that experienced analysts conduct additional secondary research before reporting a statistic.

As part of the research conducted at Iowa State University on the potential economic impacts of a biofuels ethanol plant, water, natural gas distributors, and rural electric cooperative professionals were contacted to ascertain the potential new job requirements from a large, single industry increase in demand of their respective commodities in amounts indicative of a modern 50 million gallon per year (MGY) ethanol plant. In all instances, the job requirements reported by those professionals was a tenth or less than the amount assumed in the multiplier-driven modeling systems that are commonly used (Swenson and Eathington 2006). Based on that research and on fundamental scale economy dynamics, it would not be unreasonable to assume that the marginal job gains from all new utility related activities were no greater than 25 percent of the reported values, the much lower estimates of the utility professionals notwithstanding. If that were so, and there is strong economic and practical evidence that it is, the utility job impacts could reasonably be reduced to 648 jobs leaving a total of 6,864 jobs on the operational side of ethanol and other corn processing production in Iowa.

Next to scrutinize is the reasonableness of the transportation assumptions creating 1,442 jobs. Iowa's corn historically was hauled to a mill, to a livestock feeder, or exported out of state. After processing in an ethanol refinery, the amount of 3 A Review of the Economic Rewards and Risks of Ethanol Production 63 weight that must be hauled is roughly the same as it had been when the corn was simply exported, although the nature of the haulage is changed. We can allow for a modicum of new rail capacity, new rail transport needs, and some shifting in local transportation to account for these changes; although, like the corn statistic at the start of this section, we have to conclude that nearly all of the overall transportation had already existed in the region.

Consequently, it is not unreasonable to allow for only a 25 percent bump in net new transportation jobs to the region (considering of course a substantial realignment from grain hoppers to ethanol tankers and other hauling substitutes). That would lower the 1,442 transportation jobs to 361 net new transportation jobs, thus leaving 5,782 corn processing jobs in Iowa to consider.

There are several categories of inputs that are not controvertible and would be expected to in fact be new regional indirect industrial demand linked linearly to ethanol plant operations. New ethanol plants will require substantial maintenance and repair services; they will help to stimulate demand for a variety of financial business services, to include banking, accounting, insurance, and other important activities; and they do require a new schedule of industrial chemical inputs into the production process, primarily yeasts, enzymes, and denaturants. For the time being, we can conclude that those inputs and their concomitant output and job multipliers are reasonable.

There is a fundamental question, though, about the likelihood of the bump in petroleum refinery inputs that the IRFA report claims. In all, when one looks at a modern ethanol plant's production recipe, one does not identify a set of refined petroleum product inputs (Tiffany and Eidman 2003). Their energy demands are met overwhelmingly by natural gas and electricity. The organic chemicals industry, the industry that manufactures such diverse commodities as acetone, nail polish, and tear gas along with dozens of others, however, does have strong linkages to refined petroleum products. The assumption that a modern Iowa corn ethanol dry mill operation buys $84.4 million in refined petroleum products from state suppliers as stated in the study is, however, not reasonable. It is especially dubious because Iowa's refineries made just $48.7 million in total sales across the whole state of Iowa and only needed 13 jobs to make those sales. It seems quite appropriate, then to reject the assertion that 351 refinery related jobs were created in Iowa. After all adjustments, the impact estimate has now been reduced to 5,431 total Iowa jobs that produce ethanol and other processed corn commodities, supplied non-corn inputs, or otherwise produced goods and services for the households that are supported by all of these enterprises.

The Renewable Fuels Association of Iowa report (Urbanchuck 2007) indicated that the operational side of ethanol production in Iowa ". . .support[ed] 27,200 jobs." After systematically deconstructing the authors' procedures and assumptions, however, it is more likely that somewhere around 5,431 total jobs in Iowa can be attributed to ethanol *and to all other non-fuel, corn processing production that were also counted in that analysis.* That adjusted amount is less than 20 percent of the claimed operational amount and 11.6 percent of the original grand total that included the construction jobs. It is not unreasonable to conclude that the magnitude of misstatement at the national level is often analogous to the Iowa example.

THE POLICY AND PRACTICAL IMPLICATIONS OF BLOATED ECONOMIC IMPACT CLAIMS

The foregoing assessment assists in understanding the basic job growth potential of modern ethanol production and the possible magnitude of error common in estimating that potential. The gap between perception and reality is profound and procedurally troublesome because it has implications for public policy development. Modern industrial development benefits strongly from federal, state, and local government underwriting. New ethanol plants across the U.S. are reaping large amounts of risk-reducing tax credits, subsidies, and other kinds of public support. According to one recent study (Koplow 2007), U.S. subsidies in support of ethanol production ranged from $1.42 to $1.84 per gallon in 2006 considering all capital development, credits, and other support. Using the same criteria for comparison that study concluded that subsidies for petroleum averaged just 2.4 percent of those amounts (Brasher July 2007). In Iowa, newer plants are demanding and receiving up to 20 year local property tax abatements, along with several other very valuable state tax breaks under its High Quality Job Creation Program, programs to spur capital development, and transportation assistance.

Local, state, and national public policies, incentives, and subsidies are currently allocated based on an expectation of net gains to regional economies. The IRFA study and others like it entice conclusions about the economic gains to regions that are unwarranted, however. Across the nation there is evidence of confusion and a fusion of the statistics that are used for promotion, which one must necessarily look at with a grain of salt, and of statistics that are used to justify sound public decision making, which are supposed to be based on sound scientific, economic, and policy research. If public resources are allocated on the basis of misleading or exaggerated expectations of economic gain that will not materialize, then public resources will have been squandered and the competing alternative uses to which those public resources could have been put will have been thwarted. And if so, society suffers.

ETHANOL PRODUCTION ECONOMIC OPPORTUNITIES AND OFFSETS

In a mature and relatively stable commodity production and distribution system, large changes in one segment of that system have consequences for other aspects of agriculture, non-agriculture industries, the public, and households. Initially it is important to note that the placement of a modern biofuels plant in a rural economy will result in an expansion of net regional industrial production. In the short run there is a positive economic impact to be expected. The rapid run-up in ethanol plant development in the 2005 through 2007 period, however, has also had consequences in many other aspects of agriculture, the impacts of which are just starting to be understood. This section works through some of the regional economic opportunities and offsets that must be considered as this industry matures in the Midwest.

THE INCIDENCES AND ECONOMIC BENEFITS OF FARMER OWNERSHIP ARE WANING

The majority of ethanol plants in Iowa, South Dakota, and Minnesota in the first part of this decade were considered to be "farmer" or otherwise cooperatively or locally owned. The structure of this relationship was such that corn producers as investors linked themselves to a value added production process for their commodity (Gallagher 2005). The reason for this vertical configuration was that transportation costs from some of the nation's best corn production areas ate away at much of the profits to be made from farming. The greater the production costs of shipping corn for export, for example, to the barge terminals on the Mississippi River in Minnesota, Iowa, and Illinois, the lower the price received locally. Areas with a substantial commodity price basis penalty due to transport costs had strong incentives to convert grain to more profitable uses. Livestock feeding is one value added opportunity, and ethanol production is another. A local ethanol plant allowed area farmers to receive a nominally higher price for their corn as it was not sold with the implied shipping penalty.

Most new plants are not in any meaningful sense farmer or even locally owned (Lavigne 2007). Still, there is a strong preference in the Midwest for promoting local ownership of industrial stock (Morris 2007). States like Iowa, the Dakotas, and Minnesota have, to differing degrees enacted programs and policies to promote combinations of local, often-times small or rural investors in emerging enterprises like wind energy and biofuels. The policy and development argument is that local investors will rely on local banks along with financial and legal expertise will be more likely to contract for construction and input services with local suppliers, and most of all will be likely to convert their returns on investment to local consumption and additional local investment.

While local or farmer ownership was the early model for ethanol plant development, as this industry began to rapidly grow, equity investments were sought and received from all kinds of investors from all over the country. Research at Iowa State University (Swenson and Eathington 2006) indicated that, given a 50MGY ethanol plant, the total added job impacts grew by 29 jobs for every 25 percent that the plant is owned by local residents. In short, local ownership coupled with large returns on investment locally yielded greater main street sales in the plant communities.

Those enhancements to local economic impacts were calculated based on the very robust returns received by investors in 2005 and would not be appropriate in the current market where returns are much more constrained. Importantly, those robust returns were also calculated without measuring the opportunity cost of the locally-supplied investment capital. The opportunity cost would be the normal next best alternative to which this investment money would have been put in that regional investment environment. The net return in excess of the opportunity cost is an unknown as we have no way of knowing exactly how regional investors had hitherto used their savings.

There are, therefore, three considerations that must temper the expectation of localized economic impacts from high levels of regional ownership. The value of alternative uses of that investment capital is not known, but one would assume that the normal investors' returns on all savings would have at least matched the national rate of return. Second, many farmer investors have borrowed against existing assets to invest in biofuels production. That action shifts net gains away from the now mortgaged enterprise, farming, to the new enterprise. That investment option has been widely reported, but the magnitude of it cannot be measured. Last, an increasing number of investors are not farmer-investors, and whether they reside regionally or not, there is no reason to expect those kinds of investors to behave, in the aggregate, any differently than all other investors (Lavigne 2007). Hence, for them, there is no discernible local impact to be assumed.

By the middle of 2007, growth in ethanol production capacity outstripped the national rate of absorption of ethanol and prices moderated considerably leading biofuels researchers to forecast constraints on the profitability in many of the plants, especially the older, smaller, and less efficient operations (Tokgoz et al. 2007). Consequently, one would expect that many plants are not paying substantial dividends as before, and that means the overall benefits of farmer or local ownership are expected to erode.

Higher Returns to Corn Producers and Land Owners Plus Higher Land Rents

Corn producers first promoted ethanol as a mechanism for localized gains in corn prices. The closer a corn farmer was to an ethanol plant, the better the net return on the corn as the comparatively high cost of shipping to alternative buyers was minimized. The farther a farmer was away from a plant, the less of an implied price bump (McNew and Griffith 2005). As the pace of ethanol plant expansion increased through the 2006 production year, however, corn prices nationwide, not just locally, began to climb. While corn prices demonstrate some strong fluctuations, they averaged near $2.00 for much of the previous decade. In 2006, however, average prices rose sharply as more and more plants began to process ethanol. Accordingly, the average price received nationwide rose by 58 percent over the previous year, though there is the expectation of strong localized volatility in corn prices over time as corn supplies and demand adjust (Hart 2007).

Corn farmers, however, did not see their net receipts increase by 58 percent over those two years, and in fact the U.S. BEA noted that Iowa farm earnings in 2006 were actually 5.3 percent lower than the year previous (BEA 2007) despite the corn price run-up. First, like all producers and consumers in the U.S., higher energy prices have affected farmers' bottom lines. Modern corn farming is energy intensive requiring large amounts of distillates for tractors, fertilizers derived in the main from natural gas, and propane for drying grain. So the same high oil prices boosting ethanol demand, and consequently, the demand and price received by farmers for their corn, is also boosting variable production costs on the farm. Second, as market prices increase, the total amount of government payments to corn farmers decrease, which assuredly is good news for taxpayers but still must be counted when compiling the net change in corn farmer returns and, by extension, the well being of rural economies (Westcott 2007). In all, as price increases the financial position of corn farmers improves, but the exact amount of improvement must be calculated net of subsidy reductions and the changes in all other fixed and variable costs of production changes.

Price driven gains to farmers have two very important outcomes regionally. First, they eventually help bolster the overall profitability of farming as an enterprise, which in turn is realized in higher amounts of on-farm capital and other investment along with boosted farm family spending. Second, sustained higher prices must increase the value of farm land. Over time, farmers who are landowners will realize price-induced capital gains on their land investments. For farmers that must rent their land, however, they will realize higher land use costs, which in turn will limit their net gains on production. In Iowa, according to the 2002 Census of Agriculture, 51 percent of the land in farms was rented. Higher corn prices will therefore result in increased land rent costs for 51 percent of Iowa corn crop production.

Higher Feed and Input Costs for Other Corn Consumers

Most Americans do not eat much corn. They do, on the other hand, eat a tremendous amount of products that are directly or indirectly derived from corn. Nearly all pork, beef, dairy, chicken, turkey, and egg products in the U.S. rely strongly on corn as a feedstock. Also, Americans have increasingly come to rely on high fructose corn syrups (HFCS) as a sugar substitute in many foods, beverages, and confections. It is apparent that there is strong demand for corn as a critical input into food production in the U.S.

In 2000 about 11.3 percent of all corn was made directly into food or high fructose corn syrup. Over 50 percent, however, was a feed to livestock, 16.7 percent was exported, and only 5.4 percent was used for ethanol. By 2005, the amount of feed demanded had increased to 6.1 billion bushels, but ethanol's demand for corn had increased by more than 150 percent. As a consequence of the increased demand for ethanol, the projection for 2010 has the amount of corn available for feed as eight percent lower than in 2005. At that time ethanol is expected to consume 30 percent of the nation's corn supply, up 25 percentage points in just a decade.

The high reliance on corn inputs by the livestock sector is ostensibly offset by the production of distillers' grains at the ethanol plants. Distillers' grains are the high protein residue left after the ethanol fermentation process is completed. Distillers' grains can be fed in varying degrees to livestock, ranging from 30 to 40 percent of diet to feeder cattle down to 10–20 percent for dairy cows, swine, or poultry. No matter the supply and price of distillers' grains and the mix of rations employed, feeders will still have to include some corn input costs in the mix. American cattle producers appear to be cautious about the rapid growth in the ethanol industry and have recently argued against an expansion in federal ethanol production subsidies beyond current levels (NCBA 2007), with increased corn prices as the rationale. Higher feed prices have several likely expected outcomes that may reduce meat and poultry supply. First, livestock producer net returns will shrink; this is especially the case for those that are located at some distance from ethanol plants and who had historically depended on Midwestern corn supplies. In some cases, less profitable operations will cease production entirely. In other instances, producers will not finish livestock as long – the point at which additional feed yields an optimal return will move towards a smaller animal. Hence, animals will be marketed at a lighter weight.

Finally, all consumer prices will increase as consumers absorb the higher costs associated with a lower meat and poultry supply. In all other instances, say for the production of HFCS and other corn to food products, prices will likely be passed on to consumers or otherwise result in lower returns to manufacturing producers (Westcott 2007).

In the longer term, expansion in ethanol production may lead to further concentration and vertical integration in the U.S. meat production sector. The dominant business model for poultry and meat production has a prominent firm like Tyson Foods or Smithfield Foods involved significantly with all aspects of breeding, production, processing, and distribution. As modern ethanol plants produce immense amounts of distillers' grains that are mainly suitable as cattle feed, it is possible that future ethanol plants will include very large integrated cattle feeding operations in order to efficiently feed distillers grains and to capture additional efficiencies by using animal waste as a source of fuel.

Spatial shifts in meat production are another possible outcome. Areas of the Midwest that have the highest concentration of corn production also have some of the nation's greatest concentrations of swine and poultry production because of very strong production efficiencies to be achieved from locating amidst high feed supplies. Iowa, as an example, ranks first nationally in swine and in egg production, and those animal concentrations are centered in the best corn growing areas. Cattle on feed, in large measure, are located much further to the west and southwest. Paradoxically, the animals that are least tolerant of distillers' grains and can only consume it in smaller amounts are found in higher numbers in the areas of the U.S. where there are comparatively high concentrations of ethanol plants, and the animals that are most tolerant are in comparatively lower numbers. It remains to be seen whether production advantages accumulate to the beef industry because it can more readily incorporate distillers' grains as feed and whether those advantages will work at the expense of poultry and swine production.

Grain Storage, Processing, and Distribution Systems Will Change

The nation's grain storage and transportation infrastructure developed over the years in direct response to the historical pace and pattern of crop production in the U.S. As Midwestern states have most of the nation's corn producing capacity, there are extensive systems for storage, marketing, and distributing that bounty. The nation's infrastructure for moving corn includes the inbound systems, the storage systems, grain processing systems, and the outbound systems. The nation's capacity in all aspects of managing its grain supply has developed over a long period of time and, as these are all highly capital intensive systems, that capacity closely matches production.

There are several issues affecting this complicated sector of the economy that must be taken into account as the ethanol industry develops (Ginder 2007). Ethanol plants are able to store anywhere from 10 to 25 days worth of corn. Corn that is delivered directly to the ethanol plant from farm storage, however, is corn that is not conveyed through local grain elevator systems or moved outbound via rail as historically had been the situation. So in the initial stages of ethanol plant development, gains to farmers and the expansion of ethanol production must be assessed in light of a reduction in gross receipts and reduced efficiencies on investments in all grain handling systems. As the industry matures and as competition for corn requires greater grain origination and distribution skills and efficiencies, the nation's elevator systems may come to play an integral role in moving corn into ethanol plants, but the extent and effectiveness of the sector remains to be demonstrated. In the near term, the rapid diversion of grain stocks into ethanol plants has impinged on the profitability of traditional grain handlers.

The rail transportation rolling stock that evolved to move corn is ill-suited to moving either ethanol or the byproducts of ethanol. Ethanol is primarily transported in truck and rail tankers, and cannot be transported by pipeline. Its primary byproduct is distillers' grains, which in either wet or dried form needs special rail stock as well. Furthermore, planned improvements and expansions on the Mississippi River and Illinois River locks and dams have been justified based on controversial expectations of strong growth in corn exports out of the Midwest (WSTB 2004). The expansion of ethanol production interferes with that justification in the long run, and in the short run makes the existing barge and terminal systems in the interior of the country less efficient and, therefore, less profitable.

Corn acre plantings in 2007 are estimated at 19 percent higher than 2006, and soybean plantings are down by 15 percent. Each acre of corn produces from two to three times the bushels per acre as soybeans, the primary crop sacrificed for expanded corn acres. As the nation's grain storage capacity is closely matched to grain production historical development, this rapid rise in corn supply will rapidly exhaust the nation's existing on-farm and elevator storage capacity. Storage capacity is very expensive, and it remains to be seen exactly where the economic incentives will accrue that will induce capital investment in this area. The risk, of course, is that expansion in grain storage will become potentially excess capacity if and when the nation shifts towards cellulosic ethanol production.

Spatial Changes in Crop Production

Which crop can be produced on which acre of land most profitably depends on many factors, but when the price of a commodity rises sharply, as has been the recent experience with corn in the U.S., land that had been primarily suitable for one mix of crops might now be suitable for a different mix.

Corn acreage increased in 45 of the lower 48 states between 2006 and 2007 due primarily to strong futures prices during the crop planning season of post harvest 2006 and planting time in 2007. The states of Indiana, Illinois, Minnesota, California, and North Dakota posted record corn plantings. The amount of greatest gain was in Illinois at 1.9 million more acres. A grain producing state with the strongest shift is North Dakota with nearly a 48 percent rise in corn plantings. Their increase came at the expense of a 7 percent reduction in all wheat planting and a 21 percent reduction in soybean acres. Kansas soybean plantings were down by 24 percent, Nebraska's by 21 percent, Indiana's by 19 percent, and South Dakota's by 16.5 percent. (NASS 2007).

Increased plantings of corn will affect the aforementioned storage issue: corn produces significantly more bushels per acre than either soybeans or wheat. In addition, large shifts in production will have up-stream impacts on normal regional uses of agricultural commodities. Existing processors of oil seeds for food, feed, or other uses will have sharply increased input costs due to the supply reductions. In the longer run, some commodity needs such as soybeans will necessarily be met by increased imports (Westcott 2007).

The large shift in corn acres also places stress on the nation's corn-inputs system. Corn requires fertilizers that derive mainly from natural gas, petroleum distillates for machinery, and large amounts of propane for drying corn. In all, a strong positive shift in corn production in the U.S. increases the demand for a wide array of energy inputs, which in turn drive up the prices charged to other users of those same inputs.

Finally and importantly, there are important environmental issues associated with corn production. The crop's need for high amounts of petroleum based and chemical inputs degrades groundwater and shallow aquifers. Dominant corn tilling practices also result in soil runoff, siltation of streams and rivers, and ultimately the creation of hypoxia zones in the Gulf of Mexico due to, primarily, ag-originated nutrient runoff into that area. These all entail external economic costs that are not borne by the industry or its beneficiaries, but by society at large.

There is pressure to expand the nation's land in production. There are two sources: existing pasture land and land currently enrolled in the Conservation Reserve Program (CRP). In both instances, long term land use preferences and national policy combined to remove vulnerable and marginal land from crop production. The conversion of these acres may exacerbate a wide array of environmental issues, to include increased soil erosion, surface water degradation, and soil nutrient depletion.

The Biofuels Industry will Obtain Scale Economies

Some early ethanol plants produced just 10–20 million gallons yearly (MGY) of ethanol.

Over time, ethanol plant sizes increased as investment capital became more available, as public subsidies helped to underwrite and offset risk, and as ethanol prices stabilized and demand demonstrated positive growth. Like many capital intensive industries, there are strong internal economies of scale opportunities. Economies of scale occur as a firm is able to, through more efficient utilization of its capital stock, procurement of inputs, and labor, achieve lower average costs of production per unit of output.

An obvious demonstration of scale economies presents itself readily in the ethanol industry itself. A 100 MGY per year plant only requires 46 jobs.

The plant increases its output by 100 percent, but its job needs only go up by 28.5 percent. Similarly, the plants will achieve strong efficiencies in the use of storage systems, grain moving and handling infrastructure, its land, much of its technical inputs, and larger bulk purchases of its required inputs. As the industry shifts, as firms become, on the average, larger and more efficient, larger and better operated firms, usually those that were built most recently will have higher returns per unit of production when compared to smaller and less efficient plants.

In consequence there is the expectation that in the very near future several of the nation's smaller, typically locally owned ethanol plants will become less profitable and will likely be forced out of business (Miranowski 2006).

Bioenergy Promotion and the Overall Sustainability of Rural Economies

In October of 2006 a joint U.S. Department of Energy and U.S. Department of Agriculture conference was held in St. Louis entitled

"Advanced Renewable Energy: a Rural Renaissance." New York Senator Hillary Clinton that year noted in a press release that "We can create a rural renaissance and restore the promise of Main Street. . ." in part by ". . . investing in renewable energy . . . ." (Clinton 2006). Along similar lines, U.S. Senators Norm Coleman of Minnesota and Mark Prior of Arkansas jointly proposed a *Rural Renaissance II* program in the U.S. Senate that would provide low-interest loans along with grants to rural areas to develop infrastructure and to entice investment in renewable fuels and energy sources (U.S. Senate 2005).

The head of the United Nation's Food and Agriculture Organization, Alexander Mueller, concluded in 2007 that properly promoting biofuels could be an "important tool for improving the well-being of rural people if governments take into account environmental and food security concerns." (FAO Newsroom 2007).

In each of these instances there is the assumption that the production of renewable energy from wind, corn, and biomass feedstocks will rejuvenate rural areas. Those assumptions are, however, lacking significant substantiating evidence in the near term.

For example, wind energy, which is expanding smartly in several places in the Midwest and Plains areas, is disproportionately controlled by existing, regionally dominant investor-owned utility systems. Those companies negotiate land rents for their structures, but otherwise their overall economic impact to regional economies is quite limited – once the machines are up and running, they do not require significant regionally supplied inputs.

The rural economic development potential of cellulosic systems is a complete unknown. Scientists and engineers can agree on many of the technical details and distributional requirements. Technical agreement notwithstanding, economics, however, require that the price of fuel must increase drastically before biomass can be efficiently and competitively processed.

The only realistic contemporary laboratory for gauging the revitalization potential of modern biofuels is the current expansion in corn ethanol production in the U.S. and to a lesser extent biodiesel production from oil seeds (Tokgoz et al. 2007). And the market attributes of both of those examples are distorted via the range of subsidies underwriting the current pace of growth.

There are heady expectations for growth, and some recent research (Ugarte et al. 2006) has projected that the attainment of several biofuels production goals in the U.S. will by 2030 create as many as 2.4 million new production related jobs in the U.S. were the nation to produce 60 million gallons of biofuels, many of which could accrue to rural areas.

That research is probably much too enthusiastic about the potential: much of it presupposes yet to be proven technical, distributional, investment, and policy developments that would allow for the optimization of production in attaining that optimistic goal. It also projects a future national industrial structure based primarily on the contemporary economy, a dicey prospect in economic modeling. The structure of the national economy in 2030 will be very different from the structure at present.

Putting Biofuels Job Change and Growth into Perspective in the Near Term

The interior economy of the U.S., to include its more rural areas, has not grown at anywhere near the pace as the remainder of the U.S. We also know that manufacturing in the interior of the U.S. has been hard-hit over the past decade.

Ethanol production from corn is a form of chemical manufacturing. When we look at the overall value of manufacturing to any economy, two factors are paramount: the number of jobs created and, of course, the associated earnings that workers convert to household consumption. Per unit of output, ethanol requires relatively few jobs as compared to the average manufacturing firm.

The jobs produced, however, are good jobs when measured by wage and salary. There have been very strong declines in manufacturing jobs during the present decade. Nationally, between 2000 and 2005 the nation lost nearly 3 million manufacturing positions, about 18 percent were in non-metropolitan areas of the nation, areas that did not have a central city of 50,000 or more.

The chemical manufacturing industry, of which ethanol production is a subset, lost almost 100,000 jobs over the same time period. In 2005 the average earnings of a U.S. manufacturing job considering all wages, salaries, and benefits was $60,100. In the chemical manufacturing sector it was $69,150.

The firm and job growth directly associated with ethanol production in the U.S. can be readily estimated even though current detailed U.S. statistics are not available. In 2005, just over 1.6 billion bushels of corn were converted into ethanol.

Assuming that those plants generated at a maximum 2.7 gallons of ethanol per bushel (EEOE 2007), that their average size at that time nationally was 65 million gallons per year (MGY), that they operated at 115 percent of average capacity, and that each plant averaged 38 jobs, then the U.S. ethanol industry directly required 78 plants and 2,910 jobs to process 1.6 billion bushels of corn. Average pay at new U.S. ethanol plants ranged from $45,000 to $55,000 per year – substantially less than either the U.S. manufacturing average or the average for chemical manufacturing, but substantially more than the nonfarm earnings average in most rural areas.

Were the industry to grow to process just over 4.3 billion bushels of corn annually by 2010, and assuming that plants were, on average producing 2.7 gallons of ethanol per bushel of corn, were rated at 85 MGY in average capacity, produced at 120 percent of rated capacity, and had 47 jobs per plant, then the U.S. ethanol industry would require 165 plants and 7,716 jobs in 2010.

If the rural areas of the U.S. lost some 540,000 manufacturing jobs between 2000 and 2005, it is impossible to conclude that just from corn ethanol the addition of 7,716 jobs will yield a rural renaissance.

Finally, for distributional perspective, if it is assumed that two thirds of the future corn ethanol production capacity were concentrated in Iowa, Indiana, Illinois, Nebraska, and Minnesota, then there would be, on average, one plant per just over four counties, which would work out to slightly fewer than 11.5 new manufacturing jobs per county.

The Longer Term Prospects for Rural Areas from Biofuels Development

A hallmark of modern agribusiness and modern manufacturing is the persistent substitution of capital for labor. In 1970 the average farm worker in Iowa tended 200 acres of crop land. In 2005 the average Iowa farm worker tended 300 acres of crop land.

The prospect of increased biofuels production presupposes an extension if not an acceleration in the uses of mechanical and chemical inputs into agricultural production as farmers shift production to accommodate the corn ethanol industry's rapid expansion of late.

Simultaneously, the corn ethanol industry itself will expand preferring to develop highly efficient production systems closer to the 100 MGY per year range and larger, which also will require much less labor per gallon of production than is currently the industry average.

Both of these assumptions do not portend a rural economic recovery, but rather a continuation if not an acceleration of the fundamental factors undermining most rural areas in the interior of the country: limited and specialized labor demands in only a few dominant industries that are increasingly capital intensive; and production systems that require, over time, fewer and fewer regionally supplied intermediate labor inputs.

The longer term technical and policy outlook contains an expectation of ethanol production deriving significantly from cellulosic stocks, to include ultimately acres of crop land that are dedicated to perennial energy production. If such a situation were to eventuate, then there indeed may be the potential for meaningful expansion in the value of productivity in many places of the U.S. that heretofore had not prospered. Before those unhatched chickens can be counted, however, there are several very important factors that will have to be resolved.

First and foremost, given current technology, cellulosic ethanol production, even under ideal conditions, is not cost effective. The infrastructure needs for harvesting, converting, separating, transporting, and ultimately processing cellulosic feedstocks currently do not exist and can only be imagined. The production and distributional efficiencies at the plant and spatially are significantly unknown. The overall labor requirements of processing cellulosic feedstocks is not well understood in light of the current trends in the ratios of labor to all crop acres. Shifts from one form of production, as in the current corn system, to another, such as what might eventuate from energy crop production will require a reallocation of labor and machinery, but not necessarily changes that will indirectly stimulate regional growth, especially in rural households.

The distribution of crop production and processing capacity relative to regional demand will likely favor development closer to built up areas with high demand potential to minimize transport cost and maximize returns. More remote, yet potentially productive, areas of the U.S. may realize long delays in the timing of biofuels development due to distance, infrastructure, and other constraints.

Global volatility in oil prices may not stimulate the pace and pattern of investment expected to produce expected future levels of biofuels. The nation's absorption of ethanol as a fuel source will have to increase dramatically.

And finally, an energy policy and a rural development promise that depends on rain has inherent volatility. There are many important considerations associated with biofuels production and development in the United States that were not dealt with in detail in this chapter. Enterprise-level analysis of the overall costs of operation helps policy makers and decision makers understand the production characteristics of corn and alternative ethanol production and the effects of both external and internal production factors in determining the profitability of ethanol (Tiffany and Eidman 2003).

The scope and costs of ethanol subsidies are neither detailed nor assessed here, but it must be recognized that the combined public costs of ethanol production as measured in total or on a per gallon basis is high and promises to grow. Last this analysis does not look at the overall efficacy of this form of energy development vis a vis all others. It is very difficult for many economists to discern net national gains to be derived from the current biofuels policies, and in light of that we see the rationale for ethanol promotion and biofuels development shifting from economics and economic welfare to one of "enhanced national security " (Brown 2007).

There are tangible regional economic and environmental aspects to the current debate on the development of biofuels in the U.S. Some are treating the topic in a race-to-the-moon manner with a promise of technological determinism that will, ultimately, lead to substantial social payoffs and an ultimate rationality to the process. In the meantime, however, public decision makers are charged with maximizing social gains, minimizing the undesirable consequences of public action, and assuring the nation through sound policy research that the economic benefits to be achieved from the nation's biofuels initiatives do indeed outweigh the economic, social, and environmental costs of implementing them and are, on net, better than the alternatives. To date, there is precious little evidence that is so.

4

Subsidies to Ethanol

Ethanol, or ethyl alcohol used for motor fuel, has long been used as a transport fuel. In recent years, however, it has been promoted as a means to pursue a multitude of public policy goals: reduce petroleum imports; improve vehicle emissions and reduce emissions of greenhouse gases; and stimulate rural development. Annual production of ethanol for fuel in the United States has trebled since 1999 and is expected to reach almost 7 billion gallons in 2007. This growth in production has been accompanied by billions of dollars of investment in transport and distribution infrastructure. Market factors, such as rising prices for petroleum products and state bans on methyl tertiary butyl ether (MTBE), a blending agent for which ethanol is one of the few readily available substitutes, drove some of this increase. But the main driving factor has been government support, provided at every point in the supply chain and from the federal to the local level. This chapter reviews the major policy developments affecting the fuel-ethanol industry of the United States since the late 1970s, quantifies their value to the industry, and evaluates the efficacy of ethanol subsidization in achieving greenhouse gas reduction goals. We conclude that not only is total support for ethanol already substantial — $5.8–7.0 billion in 2006 — and set to rise quickly, even under existing policy settings, but its cost effectiveness is low, especially as a means to reduce greenhouse gas emissions.

INTRODUCTION

The modern U.S. ethanol industry was born subsidized. The Energy Tax Act of 1978 introduced the first major federal subsidy for ethanol, a 4 cents-per-gallon reduction in the federal excise tax on gasohol, or E10 (a blend of 10% ethanol and 90% gasoline). In that same year, the first commercial ethanol production capacity came online. Between 1980 and 1990, production capacity more than quintupled, ending the decade at around 900 million gallons per year (mgpy). Despite a slower period of growth from the late 1980s through the mid-1990s, production capacity has grown in recent years at a very fast pace over most of the last decade. According to the Renewable Fuels Association (RFA) the main ethanol trade group, production capacity increased from 1.7 billion gallons per year (bgpy) in 1999 to 7.3 bgpy at the end of 2007 (RFA, 2007a). An additional 6.2 bgpy of capacity were under construction, the vast majority of which will rely on corn (RFA, 2007b). Meanwhile, the supply side of the ethanol market is evolving towards ever larger plants, with the largest having annual capacities approaching 300 mgpy (Planet Ark, 2006). This trend will have important effects both on feedstock supply and on the market power of different portions of the supply chain.

Conversion into ethanol serves as an increasingly important outlet for the industry's main feedstock, corn. Estimates of the share of U.S. corn production used for ethanol vary, but most place it above 20% in 2007, and likely to rise above 30% within the next few years. Despite rapid growth in demand and diversion of corn into fuel, ethanol consumption for 2006 (5.4 bgpy) supplied less than 4% of the fuel used by gasoline-powered vehicles in that year.

Industry promotion of expanded purchase mandates and continued protection from imports demonstrate that producers are counting on the government to help keep production viable. Both policies were being considered by Congress in the autumn of 2007. Even more aggressive policy interventions have also been proposed, such as setting a floor price for oil in order to protect the domestic ethanol industry from low oil prices that would render ethanol uncompetitive. Clearly, in order to understand the industry, one has to understand the roll of government incentives.

This analysis draws heavily on two in-depth studies conducted for the Global Subsidies Initiative (GSI) of the International Institute for Sustainable Development (Koplow, 2006; 2007) which in turn form part of a multi-country effort by the GSI to more thoroughly characterize and quantify subsidies to biofuels production, distribution and consumption.

This chapter first describes the evolution of government support for ethanol, focusing on the major federal programs. Thereafter follows a more detailed discussion of federal and state support policies, arranged by their point of initial economic incidence. Virtually every production stage of ethanol is subsidized somewhere in the country; in many locations, producers can tap into multiple subsidies at once. Liquid biofuels have been subsidized largely on the premise that they are domestic substitutes for imported oil; that they reduce greenhouse gas (GHG) emissions; and that they encourage rural development. Critics of subsidization have argued that the production process of these fuels is itself fossil-fuel-intensive, obviating many of the benefits of growing the energy resource; and that there are less expensive options for both GHG mitigation and rural development. Although the most recent work suggests some net fossil fuel displacement when biofuels replace petroleum products, the gains remain moderate, especially for corn-based ethanol. Others strongly contest these conclusions (e.g., Patzek, 2004; Pimentel and Patzek, 2005). Importantly, as additional analysis on modeling life-cycle impacts expands the parameters of assessment to include nitrous oxide emissions from fertilization and associated land-use changes from increased biofuel production, the net benefits of using ethanol produced from dedicated starch crops are looking less positive.

The second part of this chapter provides a variety of quantitative metrics on subsidy magnitude to illustrate how much support is being provided, not only per unit of biofuel produced, but also in terms of greenhouse gas (GHG) reductions. These values are intended to help in evaluating whether other options to diversify transport fuels or mitigate climate change might be more cost-effective.

EVOLUTION OF FEDERAL POLICIES SUPPORTING LIQUID BIOFUELS

Subsidization of ethanol production at the federal level began with the Energy Tax Act of 1978. That Act granted a 4 cents-per-gallon reduction in the federal motor fuels excise tax for gasohol, a blend of 10% ethanol and 90% gasoline, also called E10. This rate translates to 40 cents per gallon of pure ethanol at the time, and is equivalent to about $1.00 per gallon in 2007 dollars. The excise tax subsidy rate was adjusted frequently over the ensuing 25 years, until it was replaced by the Volumetric Ethanol Excise Tax Credit (VEETC) in 2004. VEETC is financed by general revenues, rather than through reduced collections for highway funding as occurred with the original exemption.

The US Congress introduced additional measures to support the ethanol industry in 1980. The Energy Security Act of 1980 initiated federally insured loans for ethanol producers, and from 1980–86 alcohol production facilities could access taxexempt industrial development bonds (Gielecki et al., 2001). Also in 1980, Congress levied a supplemental import tariff of 50 cents per gallon on foreign-produced ethanol (RFA, 2005), which was increased to 60 cents in 1984 (Gielecki et al., 2001) and now stands at 54 cents.

Several states also started to subsidize ethanol around this time. Minnesota introduced a 40 cents per gallon ethanol blenders' credit in 1980 (phased out in 1997), as did North Dakota (Sullivan, 2006). A tally of state measures carried out by the Congressional Research Service two decades ago (CRS, 1986) identified incentives in place in 29 states. By 1986, state excise-tax exemptions alone were costing state treasuries over $450 million per year (in 2007 dollars) in foregone tax receipts. In 1988, federal legislation began addressing the consumption side of the alternative fuels market. The Alternative Motor Fuels Act passed that year provided credits to automakers in meeting their Corporate Average Fuel Economy (CAFE) standards when they produced cars capable of being fueled by alternative fuels (Duffield and Collins, 2006). Earning these credits did not require that the vehicles actually run on the alternative fuels, and because so few vehicles have (somewhat less than one percent of their mileage, according to a 2002 Report to Congress), the rule has been estimated to have increased domestic oil demand by 80,000 barrels a day (MacKenzie et al., 2005).

Environmental concerns have also helped improve the market position of biofuels. The Clean Air Act Amendments of 1990 mandated changes to the composition of gasoline in an effort to address two specific air-pollution problems. Reformulated gasoline was designed to help reduce ozone-forming hydrocarbons, as well as certain air toxins in motor-vehicle emissions, and was prescribed for areas of the country suffering the most-severe ozone problems. Oxygenated fuels were intended for use in the winter, in certain metropolitan and high-pollution areas, in order to reduce emissions of carbon monoxide. An oxygen-increasing additive, or oxygenate, was required to be added to these types of gasoline reformulations. However, the Amendments did not specify any particular oxygenate (of which there are several) for achieving these goals (Liban, 1997). Mandates to use ethanol for at least 30% of the oxygenates needed to meet these requirements were promulgated by the U.S. Environmental Protection Agency (EPA) in 1994 with the strong support of the ethanol industry, but they were overturned in a court challenge a year later (Johnson and Libecap, 2001).

MTBE (methyl tertiary butyl ether), a petroleum-derived additive, emerged as the oxygenate of choice, primarily because the oil industry already had more than a decade of experience using it as an octane enhancer. Then, in 2004, concerns over the carcinogenicity of MTBE and contamination of groundwater from leaky storage tanks led several key states, starting with California, New York and Connecticut, to ban the additive (Yacobucci, 2006). By early 2006, nineteen other states had banned or limited the use of MTBE. The demise of MTBE was then accelerated by the Energy Policy Act of 2005 (EPACT05). In addition to not granting MTBE producers liability protection, Congress decided that the oxygenate mandates had yielded mediocre results, and so ended them. Effective 6 May 2006, non-oxygenated reformulated gasoline could be sold in most parts of the country (Yacobucci, 2006). With MTBE effectively no longer an option, ethanol remains as the main surviving competing fuel additive for increasing octane, a position that has helped further boost demand for the fuel.

More significantly, EPACT05 also included the first federal purchase mandates for liquid biofuels. Referred to as the "Renewable Fuels Standard" (RFS), it fixed minimum consumption levels of particular specified fuels for each year, with the mandated level rising over time. Most of the mandated volumes under present law are expected to be fulfilled by ethanol from corn.

CURRENT POLICIES SUPPORTING ETHANOL

Using a standard economic classification scheme for industry support, we provide an overview of the many types of incentives now in place to support the ethanol industry. As we were able to identify more than 200 support measures benefitting ethanol nationwide in 2006 (some of which also cover biodiesel, which is not discussed here), this section provides illustrations rather than a catalog.

VOLUME-LINKED SUPPORT

Volume-linked support takes two main forms. The first, market price support, includes interventions such as import tariffs or purchase mandates that are linked to fuel volumes but operate by raising the price received by commodity producers above what it would be in the absence of such interventions. The second includes direct payments to producers that are linked to their levels of production. In the United States, output-related subsidies for ethanol are generally linked to gallons of fuel produced or blended.

Market Price Support Associated with Tariffs and Mandates

Market price support (MPS) refers to financial transfers to producers from consumers arising from policy measures that support production by creating a gap between domestic market prices and border prices of the commodity (OECD, 2001). It can be considered the residual support element resulting from the interaction of any number of policies. Three policies play a significant role in supporting market prices for biofuels in the United States: tariffs, blending mandates, and tax credits and exemptions (de Gorter and Just, 2007). Ideally, MPS is measured by comparing actual prices obtained in a market with an appropriate reference price. Because the nature of the information on tax credits is much more concrete than that available on prices, for the purpose of this exercise we treat tax credits separately from the effects of tariffs and blending mandates.

These latter two are described briefly below. *Tariffs* — Imported fuel ethanol is currently subject to both the normal *ad valorem* tariff and a specific-rate tariff. The applied MFN (most-favored nation) tariff on imports of undenatured ethyl alcohol (80% volume alcohol or higher) is 2.5%, and on denatured ethyl alcohol it is 1.9%. The specific-rate tariff is 54 cents per gallon. Hartley (2006) notes that the supplemental tariff is punitive, since it is applied volumetrically to the full mixture (i.e., including the denaturant), and is actually higher than the domestic subsidy it supposedly offsets.

Not all ethanol imported to the United States is subject to these tariffs, however.7 Canada and Mexico—the United States' partners in the North American Free Trade Agreement (NAFTA) — for example, can export ethanol to the United States dutyfree. Countries that are covered by the Caribbean Basin Economic Recovery Act (CBERA) can export an unlimited amount of ethanol to the United States duty-free if it is made predominantly from local feedstocks, or a volume equivalent of up to seven percent of U.S. fuel-ethanol consumption if it is made mainly from feedstocks grown outside of the region (Etter and Millman, 2007).

Renewable fuels standards — As noted above, federal RFS targets of 4 bgpy in 2006, rising to 7.5 bgpy by 2012, were introduced by EPACT. Post-2012 increases are meant to occur at the same growth rate as for gasoline demand. Higher credits (equal to 2.5 times those for sugar- or starch-based ethanol) are available for cellulosic ethanol until 2012, after which 250 mgpy of cellulosic ethanol usage becomes mandatory (Duffield and Collins, 2006). Biodiesel is included at a higher credit rate as well (1.5 times that of corn ethanol) because of its higher heat rate (EPA, 2006b).

Several states have issued mandates of their own; they are often more stringent than the federal one. Minnesota had already established a renewable fuels mandate prior to the federal RFS; it requires that gasoline sold in the state must contain 20% ethanol by 2013. However, many other states have become active as well. In 2006, Iowa set a target to replace 25% of all petroleum used in the formulation of gasoline with biofuels (biodiesel or ethanol). Hawaii wants 10% of highway fuel use to be provided by alternative fuels by 2010; 15% by 2015; and 20% by 2020. A few other states have set more modest requirements, some of which (as forMontana and Louisiana) are contingent on production of ethanol within these states reaching certain minimum levels.

The combined effects of tariffs in the presence of renewable fuel standards — The main effect of a tariff is to protect domestic markets from competition from lower-priced imports, thus allowing domestic prices to rise higher than they would otherwise. When only a tariff is in place, competition from foreign suppliers of ethanol will be reduced, but domestic manufacturers must still compete with nonethanol alternatives, notably gasoline.8 Mandating a minimum market share for a good also normally drives up its price. The size of the impact will depend on a variety of factors, including how large the mandated purchases are relative to what consumption would have been otherwise; the degree to which output of the good increases as prices rise; and whether competition from imports is allowed. With a mandate but no tariff, the amount of ethanol sold domestically would possibly be higher than otherwise, but its price would be constrained by foreign sources. A mandate plus a tariff both raises the threshold price at which foreign-sourced ethanol becomes competitive, and protects domestic suppliers from being undercut by the price of gasoline.

A number of parties have tried to estimate how much the RFS mandates alone, or in combination with import tariffs, increase domestic prices of biofuels. Several reach the conclusion that increases in wholesale (also known as "rack") prices would be more than offset by government subsidies, resulting in declines in pump prices. The results of both of these studies are of course sensitive to the degree to which state and federal subsidies to ethanol would be passed on to consumers, rather than absorbed into operating margins and profits of ethanol market participants.

Others have looked mainly at producer prices. Elobeid and Tokgoz (2006) (henceforth "E&T"), analyzed the impact of liberalizing ethanol trade between the United States and Brazil using a multi-market international ethanol model calibrated on 2005 market data and policies, taking the United States' renewable fuel standard and Brazil's blending mandates as givens. Were trade barriers alone to be removed (retaining the existing renewable fuel mandate of 7.5 billion gallons per year, as well as the VEETC), they estimate the average U.S. ethanol prices from 2006 to 2015 would fall by 13.6%, or $0.27 per gallon. These results provide a rough indication of the degree to which the import tariff, in the presence of the existing (EPACT05- established) renewable fuels standard, increases the cost of meeting that standard. Should the import tariff remain in place while a higher RFS is implemented (as are proposed in pending energy legislation), the MPS would be expected to rise significantly.

Estimating market price support for a commodity ideally involves calculating the gap between the average annual unit value, or price, of the good (usually measured at the factory gate) with a reference price, usually either an average (pre-tariff) unit import price or the export price.12 Since such data are not readily available for the U.S. market, we have used the E&T results to obtain a rough estimate of market price support exclusive of the effect of the VEETC, the subsidy value of which we treat separately. Applying the E&T's price mark-up to domestically-produced ethanol generates an estimate of the contribution of the tariff to MPS of $1.3 billion in 2006, rising to more than $3 billion per year as domestic production grows.

Tax Credits and Exemptions

The federal Volumetric Ethanol Excise Tax Credit (VEETC), enacted in 2004 by the Jumpstart Our Business Strength (JOBS) Act, constitutes the single largest subsidy to ethanol. It provides a credit against income tax of 51 cents per gallon of ethanol blended into motor fuel. It is awarded without limit, and regardless of the price of gasoline, to every gallon of ethanol — domestic or imported — blended in the marketplace. Moreover, it is not subject to corporate income tax, which means its value to recipients is greater than if it were a simple grant, or a price benefit provided through an exception from an excise tax.

The Benefit of Tax Exemption for the VEETC

Tax breaks allow larger than normal deductions from taxable income or reductions in taxes due. A side-effect of the reduced tax payments is that the remaining revenues of the enterprise rise. Although the tax burden will remain lower than before the tax break, a portion of the benefit is lost to the recipient because there is some tax due on the increase in earnings. For example, under standard rules if a firm gets a $1 production tax credit (PTC), their taxes paid go down by $1, but their bottom line—which is taxable—rises by that same $1 amount. If they pay taxes at a 30% rate, they would see their taxes rise by 30 cents, leaving them with only 70 cents of the original PTC. To generate $1 in *after-tax* value to a firm, a revenue-based subsidy would need to be higher than $1 —basically $1/(1-marginal tax rate), or $1.43 in this example. This higher value is referred to as the *outlay equivalent* value of tax breaks. It was routinely reported in US tax expenditure budgets until a couple of years ago.

The question of whether a tax subsidy is exempt from taxation matters quite a bit to evaluating the distortions in energy markets from government programs. Because the VEETC is an excise tax credit rather than a production tax credit it falls into a gray area of the tax code. This ambiguity illustrates how tiny changes in the interpretation of the tax code can increase the value of subsidies to the ethanol industry by billions of dollars per year. From a technical perspective, Section 87 of the tax code specifically requires that tax credits for biofuels under Section 40 (the income tax credits) be included in taxable income, rendering their outlay equivalent value identical to the revenue loss. The language on the VEETC is not clear, however. Section 6426 of the Internal Revenue Code, which describes the VEETC, makes numerous cross-references to Section 40, mostly for definitional issues. There is no mention of Section 87.

In January of 2005, the Internal Revenue Service issued a guidance document on implementation issues related to the VEETC (IRS, 2005). Because this guidance was silent on the tax treatment of the credits, a consortium of industry groups filed comments requesting a clarification on the issue (Herman, 2005). The wording of their request indicates their inclination to treat the VEETC as not includible in taxable income until clearly instructed otherwise: Sources within both the Joint Committee on Taxation of the U.S. Congress (JCT) and the U.S. Department of Treasury have confirmed that, as of September 2007 at least, there had been no technical corrections in how the excise tax credits are treated by the Internal Revenue Service (IRS), implying that the credits are still excludible from taxable income. The incremental benefit of this exemption was roughly $1.2 billion for ethanol in 2006 on top of a direct revenue loss of $2.8 billion. The incremental subsidy from this tax loophole, supposedly a policy accident, has become the third-largest subsidy to ethanol. By 2015, even if there is no increase in the RFS, the VEETC will generate subsidies of $6.3 billion per year on a revenue loss basis and $8.9 billion per year on an outlay-equivalent basis.

In addition to the federal VEETC, several states provide reductions or exemptions for ethanol from motor fuel excise or sales taxes. The largest subsidies from these programs appear to be in Hawaii, Illinois, Indiana, and Iowa. With ethanol blends of 10% or less widely used in the country, reduced fuel taxes on E10 are becoming increasingly uncommon. Many still provide reduced rates for E85, however, and these can be fairly large per gallon. Based on the states we quantified, the average exemption for E85 was 11.5 cents per gallon; the median exemption was 7 cents per gallon. For now, the amount of ethanol consumed in E85 is small — less than 15 million gallons in 2006 according to the EIA. This is equivalent to roughly 17.4 million gallons of E85, assuming an 85% blend rate.14 The largest revenue losses tend to come from states that exempt particular fuel blends from *sales* taxes on fuels. The standard reporting of fuel tax rates provides greater clarity on deviations in excise tax rates than for fuel sales taxes. This may be one explanation for the political preference to subsidize via the sales tax. State motor-fuel tax preferences, along with state-level mandates, seem to exert a big influence on where U.S.-produced ethanol ends up being sold.

PAYMENTS BASED ON CURRENT OUTPUT

Production payments or tax credits to producers of ethanol have been on offer by the federal government and many states. These programs are normally structured to provide a pre-specified payment or tax credits for each unit (usually gallon) of output a plant produces. Supplier refunds also exist in a number of places, and operate in a similar manner.

At the federal level, the Small Producer Tax Credit, introduced in 1990, grants ethanol and biodiesel plants that produce less than

60 mgpy a 10-cents-per-gallon income-tax credit on the first 15 million gallons they produce (a maximum of $1.5 million per plant each year). Using industry data on plant nameplate capacity, we estimate the revenue loss from this provision to be over $100 million per year for ethanol. However, newer plants tend to be larger and we expect that by the end of 2009 less than 60% of the nation's ethanol plants will meet the 60 mgpy cutoff. Subsidies likely will not fall, however. When a similar situation occurred only five years ago (at which point less than 40% of the plants fell under the then 30 mgpy limit), Congress simply increased the limit.

Output-linked payments via the USDA's Bioenergy Program until recently paid an additional bounty per gallon of ethanol or biodiesel produced, with higher bounties for new production. These operated through grants rather than tax credits, but were otherwise fairly similar in structure and impact. Several states also provide production payments or tax credits for producers. Some of the programs require eligible plants to pre-qualify with the government before they can claim a credit. Some cap the total payouts (or allowable tax credits) per year to all plants. This means that the early plants may absorb the entire available funds, or that the actual per-gallon subsidy received is well below the rate nominally noted in the statute.

SUBSIDIES TO FACTORS OF PRODUCTION

Value-adding factors in biofuel production include capital, labor, land and other natural resources. Surprisingly, even labor related to biofuels production does not escape subsidization. The state of Washington, for example, allows labor employed to build biofuels production capacity, or to make biodiesel or biodiesel feedstock, to pay a reduced rate on the state's business and occupation tax.

Support for Capital Used in Manufacturing Biofuels

Scores of incentive programs have been targeted at reducing the capital cost of ethanol plants. Many of these are specific to ethanol (or ethanol and biodiesel), though others are open to a broader variety of alternative fuels. Government subsidies are often directed to encourage capital formation in a specific segment of the supply chain.

Generic Subsidies to Capital

The ethanol sector benefits from a number of important general subsidies to capital formation. Though available to a wide variety of sectors, these policies can nonetheless distort energy markets. All of them subsidize capital-intensive energy production more heavily than less capital-intensive methods. As a result, they tend to diminish the value of energy conservation relative to supply expansions. In addition, the small print in how they are defined can generate differential subsidies by sector.

Depreciation governs the process by which investments into long-lived equipment can be deducted from taxable income. The theoretical goal of depreciation is to match the cost of an asset with the period over which it will produce income, generating an accurate picture of the economics of an industry. Politically, however, depreciation schedules have become another lever used by Congress to subsidize targeted groups. Federal legislation regularly reclassifies specific industries, or shortens the period over which capital investments can be deducted from taxable income for particular sectors. This generates more rapid tax deductions. Due to the time value of money, rapid tax reductions are more valuable than those occurring slowly over time.

Production equipment for ethanol (and biodiesel) is classified as waste reduction and resource recovery plant (Class 49.5) under the Modified Accelerated Cost Recovery System (MACRS). This grouping includes "assets used in the conversion of refuse or other solid waste or biomass to heat or to a solid, liquid, or gaseous fuel," and allows full deduction of plant equipment in only seven years. An additional benefit comes in the form of the highly accelerated 200% declining balance method that can be used for Class 49.5, and that further front-loads deductions into the first years of plant operation.

With over $18 billion invested in ethanol production capacity since 2000 alone, this can constitute a fairly large subsidy. Note that our estimates incorporate only investments into plant capacity. For simplicity, we have not made similar calculations for investments in distribution infrastructure. These investments include terminals, retail facilities, tank trucks, rail cars and barges. During this same period, the ethanol industry's estimated additional spending on infrastructure assets was roughly $1 billion.

Subsidies for Specific Production-Related Capital In addition to general subsidies to capital that benefit multiple sectors of the economy, a number of subsidies target biofuel capital directly. Capital grants are used in many states and help finance production facilities, refueling or blending infrastructure, or the purchase of more expensive alternative fueled vehicles. Partial government funding of demonstration projects in the ethanol sector is common. The Energy Policy Act of 2005, for example, provided earmarked funds for a number of large biofuel-demonstration projects.

Credit subsidies, such as loans, guarantees, and access to tax-exempt debt, are common methods to subsidize the development of ethanol production and infrastructure. Title XVII of EPACT, for example, will guarantee up to 80% of the cost of selected new plants. Liquid biofuels comprised $2.5 billion of the initial round of requests for federal guarantees (DOE, 2007a), and the largest share (6 of 16) of projects chosen by the DOE to submit final funding proposals (DOE, 2007b). Program structures such as this leave little investment risk borne by investors and increase the chances of both poor project selection and of loan defaults.Many of the ethanol loan guarantees issued in the 1980s defaulted.

Some states (e.g., Delaware's Green Energy Fund) provide direct credit subsidies that are open to ethanol production facilities. Others apply their limited allowances to issue tax-exempt bonds to ethanol projects. Hawaii has authorized $50 million of tax-exempt bonds to fund a bagasse-fed ethanol plant, for example. Nebraska has authorized public power districts to build ethanol plants, and to use tax-exempt municipal bonds to finance their construction.18 New Jersey is another example, having approved $84 million in tax-exempt financing for a privately-owned ethanol plant.

Special tax exemptions for purchasing biofuels-related equipment are also common. Generally, the tax exemptions are not contingent on production levels. For example, Montana exempts all equipment and tools used to produce ethanol from grain from property taxes for a period of 10 years. In Oregon, ethanol plants pay a reduced rate (50% of statute) on the assessed value of their plant for a period of five years. These policies reduce the private cost to build a biofuels facility.

SUBSIDY STACKING

Subsidy stacking refers to a practice whereby a single plant will tap into multiple subsidy programs. This is common during the construction of a new plant, but unfortunately is often quite difficult to see when surveying subsidies. One $71-million, 20-million-gallon-per-year ethanol plant being built in Harrison County, Ohio, for example, has been able to line up government-intermediated credit or grants from seven different federal and state sources, covering 60% of the plant's capital.

Regulatory Exemptions

The waiver of regulatory requirements normally applied to similar industrial developments, but from which ethanol has been exempted, also provide a benefit equivalent to a subsidy. These exemptions can sometimes be quite surprising given ethanol's claim to be an environmentally-friendly fuel. For example, Minnesota exempts ethanol plants (though not biodiesel) with a production capacity of less than 125 mgpy from conducting an environmental impact assessment so long as the plant will be located outside of the seven-county metropolitan area.

Less stringent regulation of pollutants from the biofuels sector can also provide a benefit to the industry, by reducing its capital or operating costs. In April 2007, the EPA reclassified ethanol fuel plants from their former grouping as "chemical process plants" into a less-regulated grouping in which firms producing ethanol for human consumption had been operating. The Agency characterized the change as one of providing "equal treatment" for all corn milling facilities (EPA, 2007b). However, the change also increased the allowable air emissions from fuel ethanol facilities substantially — from 100 tons per year to 250 tons. In addition, fugitive emissions (i.e., not from the plant stack) no longer have to be tallied in the emissions total. Finally, the plants have less stringent air permitting requirements in that they no longer have to install the Best Available Control Technology (BACT). Even an industry trade magazine (Ebert, 2007) notes that [r]egardless of the legislative tributaries that many producers will have to navigate, barring litigation, most facilities will be able to take advantage of the new rule to expand and ramp up production, to build new plants with greater capacities or to potentially switch to a different power source, such as coal.

The majority of ethanol produced in the country is for fuel purposes, not human consumption.

Policies Affecting the Cost of Intermediate Inputs: Subsidies for Feedstocks

Government policies in the United States support the use of key biofuel feedstocks indirectly, through farm subsidies. Because of the United States' dominance in the global markets for corn and soybeans, federal subsidies provided to those crops during the nine years following the passage of the 1996 Farm Bill kept their farmgate prices artificially low — by an average of, respectively, 23% below and 15% below average farm production costs, according to Starmer andWise (2007). Market prices were depressed by somewhat less than the unit value of the subsidies, though the specifics varied according to market conditions. Adding to the complexity, corn and soybean markets are linked at several points. For one, the crops are often grown on the same land, in rotation. Second, they both yield competing products, such as vegetable oils and protein feeds (in the case of corn, as a byproduct of producing ethanol). These interactions complicate the way in which subsidies operate across the biodiesel and ethanol sectors.

Corn has historically been one of the most heavily subsidized crops within the United States. The Environmental Working Group (EWG), which tracks farm subsidy payments, estimates that corn subsidies totaled nearly $42 billion between 1995 and 2004 from 12 federal programs, reaching a high of $9.4 billion per year in 2005 (Environmental Working Group, 2006; Campbell, 2006). In 2006, corn did not qualify for first installments on counter-cyclical payments because the effective prices for corn exceeded its respective target price (USDA, 2006). Nonetheless, corn growers continued to receive fixed annual payments on their 2006 harvest.

Pro-rating these values to ethanol, based on the share of supply diverted to fuel production, generates an estimate of expenditure on corn subsidies associated with ethanol production of nearly $500 million for 2006, despite the sharp decline in counter-cyclical support. As ethanol production continues to consume a larger share of the domestic corn crop, its absolute (but not per-gallon) share of corn subsidies will rise accordingly.

The linkages between energy and agricultural policy are also having effects on the environment. Already, rapid growth in demand for biofuel feedstocks, particularly corn and soybeans, is changing cropping patterns in the Midwest, leading to more frequent planting of corn in crop rotations, an increase in corn acreage at the expense of wheat, and the ploughing up of grasslands (GAO, 2007). This trend is worrying, as a growing body of evidence suggests that greater carbon sequestration can be achieved through protecting natural ecosystems than by substituting biofuels for petroleum (Righelato and Spracklen, 2007).

US corn production remains chemical-intensive. Moreover, both corn and soybeans, like all row crops, typically experience higher rates of erosion than crops like wheat. Corn production is often water-intensive as well, a problem that is being exacerbated by current trends in corn-based ethanol plants. These are expanding westward, into areas more dependent on irrigation than corn produced in the Central Midwest. Some of that expansion is into counties served by the heavily overpumped Ogallala Aquifer. In addition to corn production, the ethanol plants themselves also require significant volumes of water (Zeman, 2006; National Research Council, 2007).

SUPPORT FOR R&D ON THE PRODUCTION SIDE

Federal spending on biofuels R&D hovered between $50 and $100 million a year between 1978 and 1998 (Gielecki et al., 2001). The U.S. Office of Technology Assessment reported that direct research on ethanol within the DOE was less than $15 million per year between 1978 and 1980 (OTA, 1979). It is notable that the federal government started the Bioenergy Feedstock Development Program at Oak Ridge National Laboratory nearly 30 years ago to focus on new crops and cropping systems for energy production

(Schnepf, 2007). The program continues to operate in a similar form today.24 Ethanol-related R&D is estimated to reach $400 million per year annually by 2009 (Koplow, 2007), mainly related to cellulosic ethanol.

Subsidies Related to Consumption Numerous federal and state subsidies support investment in infrastructure used to transport, store, distribute and dispense ethanol. A separate set of policies underwrites the purchase or conversion of vehicles capable of using alternative fuels.

SUBSIDIES TO CAPITAL RELATED TO FUEL DISTRIBUTION AND DISBURSEMENT

Getting ethanol from the refinery to the fuel pump requires considerable infrastructure, separate from that used to distribute gasoline. Pure ethanol attracts moisture, which means that it cannot be transported through pipelines built to carry only petroleum products. High ethanol blends, like E85, also have to be segregated and stored in corrosion-resistant tanks, and pumped through equipment with appropriate seals and gaskets. All such investment is expensive. Since 2004, the federal government and many states have started to offer financial incentives to help defray some of those costs. Under EPACT, a refueling station can obtain a tax credit that covers 30% of eligible costs of depreciable property (i.e., excluding land) for installing tanks and equipment for E85. This is capped at $30,000 per taxable year per location, and is estimated to cost the U.S. Treasury $15–30 million per year.

At least 15 states also provide assistance to establish new E85 facilities at retail gasoline outlets, as well as to support other ethanol distribution infrastructure. The Illinois E85 Clean Energy Infrastructure Development Program, for example, provides grants worth up to 50% of the total cost for converting an existing facility (up to a maximum of $2,000 per site) to E85 operation, or for the construction of a new refueling facility (maximum grant of up to $40,000 per facility). Florida recently created a credit against the state sales and use tax, available for costs incurred between 1 July 2006 and 30 June 2010, covering 75% of all costs associated with retrofitting gasoline refueling station pumps to handle ethanol; equipment for blends as low as E10 can qualify.

SUPPORT FOR VEHICLES CAPABLE OF RUNNING ON ETHANOL

The emergence of ethanol FFVs on the market provided a means for federal and state agencies to meet federal requirements for alternative fuel vehicles (AFVs) established in the Energy Policy Act of 1992. These requirements stipulated that certain government entities purchase AFVs for specified fractions (75% in the case of new light-duty vehicles) of their fleets when purchasing new vehicles. One result of this requirement was that, over time, the federal government acquired significant numbers of ethanol FFVs. Support for privately owned FFVs is also provided by several states in the form of rebates and tax credits for purchasing AFVs, or reductions on license fees and vehicle taxes, some of which apply to ethanol FFVs.

The individual states, and even some municipalities, have also provided regulatory incentives that favor AFVs. These include: the right to drive in high-occupancy vehicle (HOV) lanes, no matter how few the number of occupants in the vehicle (Arizona, California, Georgia, Utah and Virginia); the right to park in areas designated for carpool operators (Arizona); and exemptions from emissions testing (Missouri and Nevada) or certain motor-vehicle inspection programs (Ohio). Because every state develops its own definition of what exact vehicles types may participate in their AFV incentives, it is difficult to evaluate how many of these incentives apply to ethanol-powered vehicles.

AGGREGATE SUPPORT TO ETHANOL

To develop a better sense of how all of the individual subsidy programs affect the overall environment for ethanol, we have compiled a number of aggregate measures of support. The aggregate data provide important insights into a variety of policy questions, ranging from the financial cost of the support policies to taxpayers and consumers, to estimates of the costs of achieving particular policy goals. Among arguments put forth in support of biofuel subsidies are that they help the country to diversify from fossil fuels in general, and petroleum in particular; and that they have a better environmental profile than fossil fuels. Quantification is often difficult either because the subsidy's course of action is indirect (e.g., mandated use of ethanol) or because data on the magnitude of support (especially at the state level, or with tax breaks or credit enhancements) are difficult to locate. As a result, there are inevitable gaps in our subsidy tallies. Despite not counting everything, however, the subsidy picture is striking. We estimate that total support for ethanol was $5.8 billion to $7.0 billion in 2006 and, assuming no change in the RFS, will rise sharply to $11 billion by 2008 and $14 billion by 2014. The VEETC at present is the single largest ethanol subsidy and the difference between the high and low estimate is primarily associated with the incremental benefit blenders receive from the VEETC being excludible from income taxes. We believe the high estimate is a more accurate representation of government support to ethanol than is the low estimate. Subsidies from the VTEEC were $3–4 billion in 2006, and are projected to total $34 to $48 billion over the 2006–12 period.

Total undiscounted subsidies to ethanol from 2006–2012 are estimated to fall within the range of $68 billion to $82 billion. Implementation of a higher RFS (e.g., 36 bgpy by 2022) would increase total subsidies by tens of billions of dollars per Market price support, related to the combination of high barriers to imports and domestic purchase mandates, comprises the second largest subsidy to ethanol, at $1.3 billion in 2006, rising to more than $3 billion per year by 2010. Should the RFS be increased to 36 or 60 bgpy as is being considered, market price support would become the largest subsidy element, surpassing even the VEETC. Feedstock support also remains important, despite falling countercyclical payments, as direct payments remain high and ethanol is absorbing an ever-higher share of the total corn crop.

Based on 2004–2005 patterns of fuel consumption we estimate state sales and excise tax exemptions for biofuels to generate a subsidy to ethanol of approximately $400 million in 2006. Fuel taxes change regularly. In any given quarter, at least a few states will change their rates. Similarly, different sources for this information also disagree. While many states provide generous exemptions for E85, sales information are hard to come by, making revenue-loss calculations difficult. We have prorated national E85 sales data (also a few years old) by the state share of E85 refueling stations. This approach enables us to generate a rough estimate, despite the limitation of implicitly assuming that all pumps dispense the same amount of fuel per year. Rising demand; large new incentives, such as a full exemption from state taxes for E85 in New York; larger credits in Iowa; and rapidly growing sales of both ethanol blends and E85 suggest subsidies in 2007 and 2008 will be substantially higher. State policies beyond reductions in motor-fuel taxation were quantified only for 2006, based on Koplow (2006). Had these many state supports been catalogued and quantified, the magnitude of state and county supports would be much larger than what is shown in the table.

SUBSIDY PER UNIT ENERGY OUTPUT AND AS A SHARE OF RETAIL PRICE

Estimates of total support provide only a first-level indication of the potential market distortion that the subsidies may cause. Large subsidies, spread across a very large market, can have less of an effect on market structure than much smaller aggregate subsidies focused on a small market segment.

Subsidies per unit energy produced via ethanol subsidies top $11 per GJ in all years, reaching as high as $14.50 per GJ in 2008. For the 2006–12 period, subsidies to ethanol will be equal to half or more of its projected retail price. Actual price drops for ethanol during the summer of 2007 have brought prices well below the values shown in our calculations. As of October 2007, ethanol subsidies were equal to as much as 80% of the fuel's then spot-market price of roughly $1.60 per gallon (Kment, 2007; Shirek, 2007).

SUBSIDIES PER UNIT GREENHOUSE GAS DISPLACED

A common claim by biofuels supporters is that ethanol will play an important role in facilitating the transition to a society with a low carbon footprint. To test how efficient existing policies are in getting us there, we examine the subsidy cost per metric ton of CO2-equivalent displaced, and then compare this cost with the value of carbon offsets on the world's two major climate exchanges in Chicago (CCX) and Europe (ECX).

The GHG displacement factors show a large variation across data sources. This is likely due to the complexity of the systems being modeled, but the variation forms a critical policy issue. As Kammen et al. (2007: 4) note: the indirect impacts of biofuel production, and in particular the destruction of natural habitats (e.g. rainforests, savannah, or in some cases the exploitation of 'marginal' lands which are in active use, even at reduced productivity, by a range of communities, often poorer households and individuals) to expand agricultural land, may have larger environmental impacts than the direct effects. The indirect GHG emissions of biofuels produced from productive land that could otherwise support food production may be larger than the emissions from an equal amount of fossil fuels.

For corn ethanol, researchers cannot even agree on the direction of impact. Thus, at one end of the displacement factors, GHG emissions rise rather than fall from its production. This would imply very large subsidies per metric ton of *extra* CO2- equivalent emitted ($600 per metric ton in the case of corn ethanol). The best possible case for corn-based ethanol uses the lower bound subsidy estimate and divides it by the most favorable studies showing GHG reductions over the ethanol fuel cycle. Even here, subsidies per metric ton displaced are around $300.[25] Based on historical prices for carbon offsets, this same investment could have purchased 80–130 times as much displacement on the CCX, the most appropriate benchmark for the U.S. carbon market. Even on the more expensive ECX, the subsidies could have purchased 11 metric tons of offsets.

We considered also a hypothetical case assuming the same levels of government support for ethanol, but a closed-loop production system based on short-rotation poplar (*Populus* sp.) as a cellulosic feedstock. Such a production system is believed to generate net sequestration (hence its 114% displacement value). Whether the impacts would really be so low once actual crops are produced on a large scale, move outside of their optimal range, and possibly require irrigation, is an open question. The hypothetical cellulosic-ethanol case provides better tradeoffs than for corn ethanol — $110–204 per metric ton of CO_2-equivalent displaced — but the subsidies are still high: these funds could have purchased 4–8 times the offsets on the EXC or 30–85 times on the CCX.

Comparisons with Other Countries

The United States is by no means the only country that subsidizes ethanol production and consumption. Ethanol was heavily subsidized early in the development of Brazil's industry (from 1976 through 1998; see Boddey, 1993); although production is no longer directly subsidized, domestic consumption is still favored through much lower fuel-excise taxes than those applied to gasoline, and by rules preventing private ownership of diesel-powered cars. More recently, several OECD member economies have started offering reduced excise-tax rates on ethanol used as fuel, and in some cases financial assistance for ethanol-manufacturing plants. Compared with these other countries, the United States still leads in terms of absolute support provided, though per gigajoule or litre of gasoline equivalent its subsidization rate is substantially lower than those of the EU and Switzerland, which apply much higher fuel taxes to gasoline. Measured in terms of dollars per metric ton of avoided CO_2-equivalent emissions, however, the United States falls within the range of values measured for most other OECD member economies, which in all cases are orders of magnitude higher than the prices of CO_2-equivalent offsets on the major climate exchanges, as well as current estimates of the social cost of a metric ton of CO_2 emitted.

PENDING LEGISLATION

Despite a growing awareness of both the fiscal and environmental concerns about biofuels, legislative support has not abated. As of October 2007, the most "aggressive" proposed reforms (both contained in the tax section of the 2007 Farm Bill) involve reducing the excise tax credit by 5 cents per gallon (less than 10%) once the existing mandate is reached. None of the major bills would phase out the tax credits under high oil prices (when biofuels are more competitive) or remove an existing loophole that allows claimants to exclude the tax credits from their taxable income, further increasing the cost of the provision.

Several major bills under consideration by Congress, including a large proposed Energy Bill and the 2007 Farm Bill, seek to increase levels of support for biofuels, particularly ethanol. By increasing the national mandatory consumption requirement (the

Renewable Fuels Standard), for example, lawmakers hope to reduce risks to the industry of a sustained market downturn. The Energy Bill under debate in December 2007 (H.R. 6) would mandate 36 billion gallons per year by 2022. Senate Bill 23 includes a 60 billion gallon per year target by 2030. The costs of these rules are likely to be extremely large. The Energy Information Administration recently estimated that the incremental cost of a 25% renewable fuels mandate (on par with 60 billion gallons per year of biofuels) would $130 billion per year within the fuels sector alone. This translates to a cost per metric ton of CO2-equivalent reduced of more than $115, or roughly 30 times the current cost of a carbon offset on the Chicago Climate Exchange. Costs of vehicle infrastructure and increased food prices would be extra.

While the specifics of the mandates vary, most do not take into account life-cycle environmental impacts of biofuel production chains. As a result, they may encourage expensive fuels that actually worsen GHG emissions. In addition, none provide a neutral framework within which alternative ways to wean the country from imported oil and reduce greenhouse emissions can compete on a level playing field. Such alternatives include improvements in vehicle efficiency, improved maintenance and tires, and hybrid and plug-in hybrid drive trains. To further boost ethanol consumption, proposals are also being considered to increase the allowable limits for ethanol blends in gasoline for unmodified engines (currently 10%) and improve distribution infrastructure for E85. Some proposals seek to diversify the current industry by creating specific incentives for ethanol derived from feedstocks other than corn starch, expanding support for cellulosic ethanol and widening the definition of "advanced biofuels" (a definition that in some bills put before Congress would include fossil-derived fuels, and in many includes fuels derived from sugar and sorghum). As such, the new legislation compounds the current distortions to crop markets with a host of new programs to underwrite production, harvesting, storage, and the transport of cellulosic feedstocks. Some legislation makes compliance with the Renewable Fuels Standard contingent on lowering the greenhouse gas profile of biofuels (difficult to verify given problems with existing life cycle models). However, none would similarly restrict access to the excise tax credits.

CONCLUSIONS

A rapidly-expanding production base, combined with a proliferation of policy incentives, has generated a growing level of public subsidization for the ethanol industry. Many of the existing subsidies scale linearly with production capacity or consumption levels, and the resulting rate of growth in the subsidy payments can be quite large. In addition, the subsidies do not decline as the price of gasoline rises, as is the case for some subsidies benefiting petroleum and natural gas, and for some ethanol-support programs elsewhere, such as Canada (Steenblik, 2007). Although the spiraling costs of the VEETC in particular have led to discussions and proposals for subsidy phase-outs when oil prices are high (Bantz, 2006), there are currently no constraints in place.

At some point, the expiration of existing incentives may temper the growth in subsidization, but that point is still quite a few years off. Strong political support has maintained the key subsidies to ethanol for nearly 30 years, and we anticipate that those forces will remain. In the near term, we expect subsidy levels to rise sharply. Of particular interest are higher renewable fuel mandates and the rate of growth of 85% ethanol blends (E85), for which there are a number of large state subsidies that currently apply to only a small base.

Our analysis illustrates not only that subsidies to ethanol are pervasive and large, but that they are not a particularly efficient means to achieve many of the policy objectives for which they have been justified. These subsidies are the result of many independent decisions at different levels of government, resulting in policies that are often poorly coordinated and targeted. Hundreds of government programs have been created to support virtually every stage of production and consumption relating to ethanol, from the growing of the crops that are used for feedstock to the vehicles that consume the biofuels. In many locations, producers have been able to tap into multiple sources of subsidies.

Because the bulk of subsidies are tied to output and output is increasing at doubledigit rates of growth, the cost of these programs will continue to climb. Production is subsidized at the federal level even though consumption of it is mandated through the RFS. Ethanol production is supported on the grounds that it helps wean the United States from imported petroleum, but special loopholes in vehicle efficiency standards for flexible fuel vehicles (including those that run on high ethanol blends) result in higher oil imports (MacKenzie et al., 2005). The maintenance of a high tariff on imported ethanol (2.5% plus 54 cents per gallon), in particular, sits at odds with the professed policy of the U.S. government to encourage the substitution of gasoline by ethanol.

The absolute value of the subsidies is not the only, and perhaps not the main, indicator of the market-distorting potential of a set of support policies. Subsidies as a share of market price were above 40% as of mid-2006, for example, which is high in comparison with other fuels. Such high rates of subsidization might be considered reasonable if the industry was new, and ethanol was being made on a small-scale, experimental basis using advanced technologies. But that is not the case: the vast majority of subsidized production relies on mature technologies that, notwithstanding progressive improvements, have been around for decades. Ethanol also has some greenhouse gas and local-pollution benefits. But the cost of obtaining a unit of CO_2-equivalent reduction through subsidies to the fuel is extremely high: we calculate that it comes to nearly $300 per metric ton of CO_2 removed for corn-based ethanol, even when assuming an efficient plant using lowcarbon fuels for processing. Yet even under such best-case scenario assumptions for GHG reductions from corn-based ethanol, one could have achieved far more reductions for the same amount of money by simply purchasing the reductions in the marketplace. The cost per metric ton of reductions achieved through public support of corn-based ethanol already programmed over the next several years could purchase more than 10 times the offsets on the European Climate Exchange, or nearly 90 times the offsets on the Chicago Climate Exchange.

Most importantly, the U.S. government has neglected what should be its core role: to adopt a neutral strategy equally accessible to all potential options to reduce the country's reliance on imported oil. Such a strategy would not favor ethanol, but would encourage a range of potential solutions such as more efficient vehicles, better fleet maintenance, and alternative drive-trains such as plug-in hybrids. Similarly, the government has yet to indicate an exit strategy to wean the ethanol industry from protection and subsidies. Indeed, as is often the case with subsidies, current legislative proposals appear to entrench existing arrangements. These will ensure that the biofuel industry remains a significant drain on U.S. taxpayers for decades to come; and that improvements in transport-fuel options will be both slower and more expensive than would occur with a technology-neutral approach.

5

Peak Oil, Eroi, Investments and the Economy in an Uncertain Future

The issues surrounding energy are far more important, complex and pervasive than normally considered from the perspective of conventional economics, and they will be extremely resistant to market-based, or possibly any other, resolution. We live in an era completely dominated by readily available and cheap petroleum.

This cheap petroleum is finite and currently there are no substitutes with the quality and quantity required. Of particular importance to society's past and future is that depletion is overtaking technology in many ways, so that the enormous wealth made possible by cheap petroleum is very unlikely to continue very far into the future.

What this means principally is that investments will increasingly have to be made into simply getting the energy that today we take for granted, the net economic effect being the gradual squeezing out of discretionary investments and consumption. While there are certainly partial "supply-side" solutions to these issues, principally through a focus on certain types of solar power, the magnitude of the problem will be enormous because of the scale required, the declining net energy supplies available for investment and the relatively low net energy yields of the alternatives. Given that this issue is likely to be far more immediate, and perhaps more important, than even the serious issue of global warming it is remarkable how little attention we have paid to understanding it or its consequences.

INTRODUCTION

The enormous expansion of the human population and the economies of the United States and many other nations in the past 100 years have been accompanied by, and allowed by, a commensurate expansion in the use of fossil (old) fuels, meaning coal, oil and natural gas. To many energy analysts that expansion of cheap fuel energy has been the principal enabler of economic expansion, far more important than business acumen, economic policy or ideology although they too may be important.

While we are used to thinking about the economy in monetary terms, those of us trained in the natural sciences consider it equally valid to think about the economy and economics from the perspective of the energy required to make it run. When one spends a dollar, we do not think just about the dollar bill leaving our wallet and passing to some one else's. Rather, we think that to enable that transaction, that is to generate the good or service being purchased, an average of about 8,000 kilojoules of energy (equal to roughly the amount of oil that would fill a coffee cup) must be extracted from the Earth and turned into roughly a half kilogram of carbon dioxide (U.S. Statistical Review, various years). Take the money out of the economy and it could continue to function through barter, albeit in an extremely awkward, limited and inefficient way.

Take the energy out and the economy would immediately contract immensely or stop. Cuba found this out in 1991 when the Soviet Union, facing its own oil production and political problems at that time, cut off Cuba's subsidized oil supply. Both Cuba's energy use and its GDP declined immediately by about one third, all groceries disappeared from market shelves within a week and the average Cuban lost 20 pounds (Quinn 2006). Cuba subsequently learned to live, in some ways well, on about half the oil as previously, but the impacts were enormous.While the United States has become more efficient in using energy in recent decades, most of this is due to using higher quality fuels, exporting heavy industry and switching what we call economic activity (e.g. Kaufmann 2004). Many other countries, including efficiency leader Japan, are becoming substantially less efficient (Hall and Ko, 2007, LeClerc and Hall 2007, Smil 2007, personal communication).

THE AGE OF PETROLEUM

The economy of the United States and the world is still based principally on "conventional" petroleum, meaning oil, gas and natural gas liquids. *Conventional* means those fuels derived from geologic deposits, usually found and exploited using drill bit technology, and that move to the surface because of their own pressure or with pumping or additional pressure supplied by injecting natural gas, water or occasionally other substances into the reservoir. *Unconventional* petroleum includes shale oil, tar sands and other bitumens usually mined as solids and also coal bed and certain other methane deposits. For the economies of both the U.S. and the world nearly two thirds of our energy comes from conventional petroleum, about 40 percent from conventional liquid petroleum and another 20–25 percent from gaseous petroleum. Coal, and natural gas provide most of the rest of the energy that we use. Hydroelectric power and wood together are renewable energies generated from current solar input and provide about five percent of the energy that the US uses. "New renewables" including windmills and photovoltaics, provide less than one percent, and are not growing as rapidly in magnitude globally (although they are as a percent of their own contribution) as petroleum. Thus the annual increase in oil and gas use is much greater than the new quantities coming from the new renewables, at least to date. All of these proportions have not changed very much since the 1970s in the United States or the world. We believe it most accurate to consider the times that we live in as the age of petroleum, for petroleum is the foundation of our economies and our lives. Just look around.

Petroleum is especially important because of its magnitude of current use, because it has important and unique qualitative attributes leading to high economic utility that include very high energy density and transportability (Cleveland 2005), and because its future supply is worrisome. The issue is not the point at which oil actually runs out but rather the relation between supply and potential demand. Barring a massive worldwide recession demand will continue to increase as human populations, petroleum-based agriculture and economies (especially Asian) continue to grow. Petroleum supplies have been growing most years since 1900 at two or three percent per year, a trend that most investigators think cannot continue (e.g. Campbell and Laherrere 1998, Heinberg 2003). Peak oil, that is the time at which an oil field, a nation or the entire world reaches its maximum oil production and then declines, is not some abstract issue debated by theoretical scientists or worried citizens but an actuality that occurred in the United States in 1970 and in some 60 (of 80) other oil-producing nations since (Hubbert 1974, Strahan 2007, Energyfiles 2007). Several prominent geologists have suggested that it may have occurred already for the world, although that is not clear yet (e.g. Deffeyes 2005, see EIA 2007, IEA 2007).

With global demand showing no sign of abating at some time it will not be possible to continue to increase petroleum supplies, especially oil globally and natural gas in North America, or even to maintain current levels of supply, regardless of technology or price. At this point we will enter the second half of the age of oil (Campbell 2005). The first half was one of year by year growth, the second half will be of continued importance but year by year decline in supply, with possibly an "undulating plateau" at the top and some help from still-abundant natural gas outside North America separating the two halves and buffering the impact somewhat for a decade or so. We are of the opinion that it will not be possible to fill in the growing gap between supply and demand of conventional oil with e.g. liquid biomass alternatives on the scale required (Hall et al. in review), and even were that possible that the investments and time required to do so would mean that we needed to get started some decades ago (Hirsch et al. 2005). When the decline in global oil production begins we will see the "end of cheap oil" and a very different economic climate.

The very large use of fossil fuels in the United States means that each of us has the equivalent of 60–80 hard working laborers to "hew our wood and haul our water" as well as to grow, transport and cook our food, make, transport and import our consumer goods, provide sophisticated medical and health services, visit our relatives and take vacations in far away or even relatively near by places. Simply to grow our food requires the energy of about a gallon of oil per person per day, and if a North American takes a hot shower in the morning he or she will have already used far more energy than probably two thirds of the Earth's human population use in an entire day. Oil is especially important for the transportation of ourselves and of our goods and services, and gas for heating, cooking, some industries and as a feedstock for fertilizers and plastics.

How much Oil will we be able to Extract?

So the next important question is how much oil and gas are left in the world? The answer is a lot, although probably not a lot relative to our increasing needs, and maybe not a lot that we can afford to exploit with a large financial and, especially, energy profit. We will probably always have enough oil to oil our bicycle chains. The question is whether we will have anything like the quantity that we use now at the prices that allow the things we are used to having. Usually the issue of how much oil remains is not developed from the perspective of "when will we run out" but rather "when will we reach 'peak oil' globally". World wide we have consumed a little over one trillion barrels of oil. The current debate is fundamentally about whether there are 1, 2 or even 3.5 trillion barrels of economically extractable oil left to consume. Fundamental to this debate, yet mostly ignored, is an understanding of the capital, operating and environmental costs, in terms of money and energy, to find, extract and use whatever new sources of oil remain to be discovered, and to generate whatever alternatives we might choose to develop. Thus the investment issues, in terms of both money and energy, will become ever more important.

There are two distinct camps for this issue. One camp, which we call the "technological cornucopians", led principally by economists such as Michael Lynch (e.g. Lynch 1996, Adelman and Lynch 1997), believes that market forces and technology will continue to supply (at a price) more or less whatever oil we need for the indefinite future. They focus on the fact that we now are able to extract only some 35 percent of the oil from an oil field, that large areas of the world (deep ocean, Greenland, Antarctica) have not been explored and may have substantial supplies of oil, and that substitutes, such as oil shale and tar sands, abound. They are buoyed by the failure of many earlier predictions of the demise or peak of oil, two recent and prestigious analyses by the U.S.

Geological Survey and the Cambridge Energy Research Associates that tend to suggest that remaining extractable oil is near the high end given above, the recent discovery of the deepwater Jack 2 well in the Gulf of Mexico and the development of the Alberta Tar Sands, which are said to contain more oil than remains even in Saudi Arabia. They have a strong faith in technology to increase massively the proportion of oil that can be extracted from a given oil field, believe that many additional fields await additional exploration, and believe there are good substitutes for oil.

A second camp, which we can call the "peak oilers", is composed principally of scientists from a diversity of fields inspired by the pioneering work of M. King Hubbert (e.g. 1969; 1974), a few very knowledgeable and articulate politicians such as US Representative Roscoe Bartlett of Maryland, many private citizens from all walks of life and, increasingly, some members of the investment community. All believe that there remains only about one additional trillion barrels of extractable conventional oil and that the global peak – or perhaps a "bumpy plateau", in extraction will occur soon, or, perhaps, has already occurred. The arguments of these people and their organization, the Association for the Study of Peak Oil (ASPO), spearheaded by the analyses and writings of geologists Colin Campbell and Jean Laherrere, are supported by the many other geologists who more or less agree with them, the many peaks that have already occurred for many dozens of oil-producing countries, the recent collapse of production from some of our most important oil fields and the dismal record of oil discovery since the 1960s – so that we now extract and use four or five barrels of oil for each new barrel discovered. They also believe that essentially all regions of the Earth favourable for oil production have been well explored for oil, so that there are few surprises left except perhaps in regions that will be nearly impossible to exploit.

There are several issues that tend to muddy the water around the issue of peak oil. First of all, some people do, and some do not, include natural gas liquids or condensate (liquid hydrocarbons that condense out of natural gas when it is held in surface tanks). These can be refined readily into motor fuel and other uses so that many investigators think they should simply be lumped with oil, which most usually they are. Since a peak in global natural gas production is thought to be one or two decades after a peak in global oil, inclusion of natural gas liquids extends the time or duration of whatever oil peak may occur. Consequently, if indeed peak oil has occurred, a peak in liquid petroleum fuels might still be before us. A second main issue is "how much oil is likely ever to be produced" vs. "when will global production peak, or at least cease growing?"

In theory the issues are linked, perhaps tightly, but it is probably far more important to focus on the peak production rate rather than the total quantity that we will ever extract. In terms of ultimate economic impact, and probably prices, the most important issue is almost certainly the ratio between the production rate and its increase or decrease, and the consumption rate and its increase or decrease. Both the production and the consumption of oil and also natural gas have been growing at roughly two percent a year up through at least 2006. The great expansion of the economies of China and India, which at this time show no evidence of a slowdown, have recently more than compensated for some reduced use in other parts of the world. Nevertheless the growth rate of the human population has been even greater so that "per capita peak oil" probably occurred in 1978 (Duncan 2000).What the future holds may have more to do with the consumption rate than the production rate. If and when peak petroleum extraction occurs it is likely to increase prices which should bring an economic slowdown which should decrease oil use which might decrease prices and . . . the chickens and eggs can keep going for some time. That is why many peak oilers speak of "a bumpy plateau".

However if potential demand keeps growing then the difference between a steady or declining supply and an increasing demand presumably would continue upward pressures on price.

The rates of oil and gas production (more accurately extraction) and the onset of peak oil are dependent upon many interacting factors, including geological, economic and political. The geological restrictions are the most absolute and depend on the number and physical capacity of the world's operating wells. In most fields the oil does not exist in the familiar liquid state but in what is more akin to a complex oil-soaked brick. The rate at which oil can flow through these "aquifers" depends principally upon the physical properties of the oil itself and of the geological substrate, but also upon the pressure behind the oil that is provided initially by the gas in the well. Then, as the field matures, the pressure necessary to force the oil through the substrate to the collecting wells is supplied increasingly by pumping more gas or water into the structure. As with water wells the more rapidly the oil is extracted the more likely the substrate will become compacted, restricting future yields. Detergents, CO_2 and steam can increase yields but too-rapid extraction can cause compaction of the "aquifer" or fragmentation of flows which reduce yields. So our physical capacity to produce oil depends upon our ability to keep finding large oil fields in regions that we can reasonably access, our willingness to invest in exploration and development, and our willingness to not produce too quickly. The usual economic argument is that if supply is reduced relative to demand then the price will increase which will then signal oil companies to drill more, leading to the discovery of more oil and then additional supply. Although that sounds logical the results from the oil industry might not be in accordance to that logic as the empirical record shows that the rate at which oil and gas is found has little to do with the rate of drilling.

It is thought that at this time we are producing oil globally pretty nearly to our present capacity, although future depletion or new fields can change that. Finally, output can be limited or (at least in the past) enhanced for political reasons – which are even more difficult to predict than the geological restrictions. Empirically there is a fair amount of evidence from post peak countries, such as the U.S., that the physical limitations become important when about half of the ultimately-recoverable oil has been extracted. But why should that be? In the US it certainly was not due to a lack of investment, since most geologists believe that the US had been over drilled. We probably will not know until we have much more data, and much of the data are closely guarded industry or state secrets. According to one analyst if one looks at all of the 60 or so post peak oil-producing countries the peak occurs on average when about 54 percent of the total extractable oil in place has been extracted (Energyfiles. com 2007). Finally oil-producing nations often have high population and economic growth, and are using an increasing proportion of their own production (Hallock et al. 2004).

The United States clearly has experienced "peak oil". In a way this is quite remarkable, because as the price of oil increased by a factor of ten, from 3.50 to 35 dollars a barrel during the 1970s, a huge amount of capital was invested in US oil discovery and production efforts so that the drilling rate increase from 120 million feet per year in 1970 to 400 million feet in 1985. Nevertheless the production of crude oil decreased during the same period from the peak of 3.52 billion barrels a year in 1970 to 3.27 in 1985 and has continued to decline to 1.89 in 2005 even with the addition of Alaskan production. Natural gas production has also peaked and declined, although less regularly. Thus despite advancement of petroleum discovery and production technology, and despite very significant investment, U.S. production has continued its downward trend since 1970. The technological optimists are correct in saying that advancing technology is important. But there are two fundamental and contradictory forces operating here, technological advances and depletion. In the US oil industry it is clear that depletion is trumping technological progress, as oil production is declining and oil is becoming much more expensive to produce.

DECREASING ENERGY RETURN ON INVESTMENT

Energy return on investment (EROI or EROEI) is simply the energy that one obtains from an activity compared to the energy it took to generate that energy. The procedures are generally straightforward, although rather too dependent upon assumptions made as to the boundaries, and when the numerator and denominator are derived in the same units, as they should, it does not matter if the units are barrels (of oil) per barrel, Kcals per Kcal or MJoules per Mjoule as the results are in a unitless ratio. The running average EROI for the finding and production of US domestic oil has dropped from greater than 100 kilojoule returned per kilojoule invested in the 1930s to about thirty to one in the 1970s to between 11 and 18 to one today. This is a consequence of the decreasing energy returns as oil reservoirs are increasingly depleted and as there are increases in the energy costs as exploration and development are shifted increasingly deeper and offshore (Cleveland et al. 1984, Hall et al. 1986, Cleveland 2005). Even that ratio reflects mostly pumping out oil fields that are half a century or more old since we are finding few significant new fields. (In other words we can say that new oil is becoming increasingly more costly, in terms of dollars and energy, to find and extract). The increasing energy cost of a marginal barrel of oil or gas is one of the factors behind their increasing dollar cost, although if one corrects for general inflation the price of oil has increased only a moderate amount until 2007.

The same pattern of declining energy return on energy investment appears to be true for global petroleum production. Getting such information is very difficult, but with help from the superb database of the John H. Herold Company, several of their personnel, and graduate student and sometime Herold employee Nate Gagnon we were able to generate an approximate value for global EROI for finding new oil and natural gas (considered together). Our preliminary results indicate that the EROI for global oil and gas (at least for that which was publically traded) was roughly 26:1 in 1992, increased to about 35:1 in 1999, and since has fallen to approximately 19:1 in 2005. The apparent increase in EROI during the late 1990s is during a period when drilling effort was relatively low and may reflect the effects of reduced drilling effort as was seen for oil and gas in the United States. If the rate of decline continues linearly for several decades then it would take the energy in a barrel of oil to get a new barrel of oil.

While we do not know whether that extrapolation is accurate, essentially all EROI studies of our principal fossil fuels do indicate that their EROI is declining over time, and that EROI declines especially rapidly with increased exploitation (e.g. drilling) rates. This decline appears to be reflected in economic results. In November of 2004 The New York Times reported that for the previous three years oil exploration companies worldwide had spent more money in exploration than they had recovered in the dollar value of reserves found. Thus even though the EROI of global oil and gas is still about 18:1 as of 2006, this ratio is for all exploration and production activities. It is possible that the energy break even point has been approached or even reached for finding new oil. Whether we have reached this point or not the concept of EROI declining toward 1:1 makes irrelevant the reports of several oil analysts who believe that we may have substantially more oil left in the world, because it does not make sense to extract oil, at least for a fuel, when it requires more energy for the extraction than is found in the oil extracted.

How well we weather this coming storm will depend in large part on how we manage our investments now. From the perspective of energy there are three general types of investments that we make in society. The first is investments into getting energy itself, the second is investments for maintenance of, and replacing, existing infrastructure, and the third is discretionary expansion. In other words before we can think about expanding the economy we must first make the investments into getting the energy necessary to operate the existing economy, and into maintaining the infrastructure that we have, at least unless we wish to accept the entropy-driven degradation of what we already have. Investors must accept the fact that the required investments into the second and especially the first category are likely to increasingly limit what is available for the third. In other words the dollar and energy investments needed to get the energy needed to allow the rest of the economy to operate and grow have been very small historically, but this is likely to change dramatically. This is true whether we seek to continue our reliance on ever-scarcer petroleum or whether we attempt to develop some alternative. Technological improvements, if indeed they are possible, are extremely unlikely to bring back the low investments in energy that we have grown accustomed to.

The main problem that we face is a consequence of the "best first" principle. This is, quite simply, the characteristic of humans to use the highest quality resources first, be they timber, fish, soil, copper ore or, of relevance here, fossil fuels. This is because economic incentives are to exploit the highest quality, least cost (both in terms of energy and dollars) resources first, as was noted 200 years ago by economist David Ricardo (e.g. 1891). We have been exploiting fossil fuels for a long time. The peak in finding oil was in the 1930s for the United States and in the 1960s for the world, and both have declined enormously since then. An even greater decline has taken place in the efficiency with which we find oil, that is the amount of energy that we find relative to the energy we invest in seeking and exploiting it. As a consequence of the decreasing energy returns as oil depletion increases, and of the increasing energy costs as exploration and development shifted increasingly deeper offshore or into increasingly hostile environments, the energy return on investment (EROI) for US domestic oil has declined to perhaps 15 to one today, even though that contemporary ratio reflects mostly pumping out oil fields that are half a century or older. In other words we can say that new oil is becoming increasingly more costly, in terms of energy (and consequently dollars), to find and extract. The alternatives to oil available to us today are characterized by even lower EROIs, limiting their economic effectiveness. It is critical for CEOs and government officials to understand that the best oil and gas are simply gone, and there is no easy replacement.

That pattern of exploiting and depleting the best resources first also is occurring for natural gas. Natural gas was once considered a dangerous waste product of oil development and was burned or flared at the well head. But during the middle years of the last century large gas pipeline systems were developed in the U.S. and Europe that enabled gas to be sent to myriad users who increasingly discovered its qualities of ease of use and cleanliness, including its relatively low carbon dioxide emissions, at least relative to coal. US natural gas originally came from large fields in Louisiana, Texas and Oklahoma. Its production has moved increasingly to smaller fields distributed throughout Appalachia and, increasingly, the Rockies. As the largest fields that traditionally supplied the country peaked and declined a national peak in production occurred in 1973, and then as "unconventional" fields were developed a second, somewhat smaller peak occurred in 2001. Gas production has fallen by about 6 percent from that peak, and many investigators predict a "natural gas cliff" as traditional fields are exhausted and as it is increasingly difficult to bring smaller unconventional fields on line to replace the depleted giants. There had been an encouragement of electricity production from natural gas because it is relatively clean, but a large loss of petrochemical companies from the US because of the increasing price.

THE BALLOON GRAPH

We pay for imported oil in energy as well as dollars, for it takes energy to grow, manufacture or harvest what we sell abroad to gain the foreign exchange with which we buy fuel, (or we must in the future if we pay with debt today). In 1970 we gained roughly 30 megajoules for each megajoule used to make the crops, jet airplanes and so on that we exported (Hall et al. 1986). But as the price of imported oil increased, the EROI of the imported oil declined. By 1974 that ratio had dropped to nine to one, and by 1980 to three to one. The subsequent decline in the price of oil, aided by the inflation of the export products traded, eventually returned the energy terms of trade to something like it was in 1970, at least until the price of oil started to increase again after 2000. A rough estimate of the quantity and EROI of various major fuels in the U.S., including possible alternatives. An obvious aspect of that graph is that qualitatively and quantitatively alternatives to fossil fuel have a very long way to go to fill the shoes of fossil fuels. This is especially true when one considers the additional qualities of oil and gas, including energy density, ease of transport and ease of use.

The implications of all this is that if we are to supply into the future the amount of petroleum that the US consumed in the first half of this decade it will require enormous investments in either additional unconventional sources, in import facilities or as payments to foreign suppliers. That will mean a diversion of investment capital and of money more generally from other uses into getting the same amount of energy just to run the existing economy. In other words investments, from a national perspective, will be needed increasingly just to run what we have, not to generate real new growth. If we do not make these investments our energy supplies will falter or we will be tremendously beholden to foreigners, and if we do the returns may be small to the nation, although of course if the price of energy increases greatly the returns to the individual investor may be large. Another implication is if this issue is as important as we believe it is then we must pay much more attention to the quality of the data we are getting about energy costs of all things we do, including getting energy. Finally the failure of increased drilling to return more fuel calls into question the basic economic assumption that scarcity-generated higher prices will resolve that scarcity by encouraging more production. Indeed scarcity encourages more exploration and development activity, but that activity does not necessarily generate more resources. It will also encourage the development of alternative liquid fuels, but their EROIs are generally very low.

Economic Impacts of Peak Oil and Decreasing EROI

Whether global peak oil has occurred already or will not occur for some years or, conceivably, decades, its economic implications will be enormous because we have no possible substitute on the scale required and because alternatives will require enormous investments in money and energy when both are likely to be in short supply.

Despite the projected impact on our economic and business life within relatively few years or at most decades, neither government nor the business community is in any way prepared to deal with either the impacts of these changes or the new thinking needed for investment strategies.

The reasons are myriad but include: the disinterest of the media, the failure of government to fund good analytic work on the various energy options, the erosion of good energy record keeping at the department of commerce, and the lack of really good options. The second point perhaps is debatable but we stand on that statement because of the top ten or so energy analysts that we are familiar with none are supported by government, or generally any, funding.

There are not even targeted programs in NSF or the Department of Energy where one might apply if one wishes to undertake good objective, peer reviewed EROI analyses. Consequently much of what is written about energy is woefully misinformed or simply advocacy by various groups that hope to profit from various perceived alternatives.

It is not unlikely that issues pertaining to the end of cheap petroleum will be the most important challenge that Western society has ever faced, especially when considered within the context of our need to deal with climate change and other environmental issues related to energy. Any business leaders who do not understand the inevitability, seriousness and implications of the end of cheap oil, or who make poor decisions in an attempt to alleviate its impact, are likely to be tremendously and negatively impacted as a result. At the same time the investment decisions we will make in the next decades will determine whether civilization is to make it through the transition away from petroleum or not.

What would be the impacts of a large increase in the energy and dollar cost of getting our petroleum, or of any restriction in its availability? While it is extremely difficult to make any hard predictions, we do have the record of the impacts of the large oil price increases of the 1970s as a possible guide.

These "oil shocks" had very serious impacts on our economy which we have examined empirically in past publications (e.g. Hall et al. 1986).

Many economists then and now did not think that even large increases in the price of energy would affect the economy dramatically because energy costs were but three to six percent of GDP. But by 1980, following the two "oil price shocks" of the 1970s, energy costs had increased dramatically until they were 14 percent of GDP. Actual shortages would have even greater impacts, if for example sufficient petroleum to run our industries or businesses were not available at any price. Other impacts included, and would include, an enhancement of our trade imbalances as more income is diverted overseas, adding to the foreign holdings of our debt and a decrease in discretionary disposable income as more money is diverted to access energy, whether via higher prices, more petroleum exploration or low EROI alternative fuels. This in turn would affect those sectors of the economy that are not essential. Consumer discretionary spending would probably fall dramatically, greatly effecting non-essential businesses such as tourism.

THE "CHEESE SLICER" MODEL

We have attempted to put together a conceptual and computer model to help us understand what might be the most basic implications of changing EROI on the economic activity of the United States. The model was conceptualized when we examined how the U.S. economy responded to the "oil shocks" of the 1970s. The underlying foundation is the reality that the economy as a whole requires energy (and other natural resources derived from nature) to run, and without these most basic components it will cease to function.

The other premise of this model is that the economy as a whole is faced with choices in how to allocate its output in order to maintain itself and to do other things. Essentially the economy (and the collective decision makers in that economy) has opportunity costs associated with each decision it makes. The large square represents the structure of the economy as a whole, which we put inside a symbol of the Earth biosphere/geosphere to reflect the fact that the economy must operate within the biosphere (e.g. Hall et al. 2001). In addition, of course, the economy must get energy and raw materials from *outside* the economy, at least as narrowly perceived, that is from nature (i.e. the biosphere/ geosphere).

The output of the economy, normally considered GDP, is represented by the large arrow coming out of the right side, where the depth of the arrow represents 100 percent of GDP.

For the sake of developing our concept we think of the economy, for the moment, as an enormous dairy industry and cheese as the product coming out of the right hand side, moving towards the right.

This output (i.e. the entire arrow) could be represented as either money or embodied energy. We use the former in this analysis (as almost all of the relevant data is recorded in monetary, not energy, units), but it is probably not terribly different from using energy outputs.

So, our most important question is "how do we slice the cheese", that is how do we, and how will we divide up the output of the economy, or said differently, in what way can the output of the economy be divided up with the least objectionable opportunity cost. Most economists might answer "according to what the market decides," that is according to consumer tastes and buying habits. But we want to think about it a little differently because we think things might be profoundly different in the future.

Most generally the output of the model (and the economy) has two destinations: *investment* or *consumption*. These could be divided further into private vs. government investments and consumptions, or we could add in debt service, exports and imports, and so on.

We choose not to do this, at least initially, although we are co-developing a much more complicated model of the economy with the Millennium Institute called T-21 North America. (e.g. Millennium Institute 2007). Our next question is "investment or consumption of/for what?" To do this we ask about what we must spend our money/energy on, that is on the *required* expenditures (without which the economy would cease to function).

These include the investments in maintaining societal infrastructure (i.e. repairing and rebuilding bridges, roads, machines, factories, vehicles – represented by the top middle arrow feeding back from output of the economy back to the economy itself), and some kind of minimal food, shelter and clothing for the population (represented by the bottom rightward pointing arrow) required to maintain all individuals in society at the level of the Federal minimum standard of living).

Another, and very important, essential need, the one of greatest interest to us here, is the investments into, or payments for, energy (i.e. the amount of diverted economic output that is used to secure and purchase our imported oil) that is required for the energy needed for the economy to operate at any particular level. This energy is absolutely critical for the economy to operate and must be paid for through proper payments and investments – which we consider together as investments to get energy. No investment in energy, no economic output.

While there may be other critical inputs from nature (water, minerals etc.) or society (educated workforce etc.) we ignore these for the moment as they are unlikely to be as volatile as energy. This "Energy Investment" feedback is represented by the topmost arrow from the output of the economy back upstream to the "workgate" symbol (Odum 1971).

The width of this line represents the investment of energy into getting more energy. Of critical interest here is that as the EROI of our economy's total combined fuel source declines then more and more of the output of the economy *must* be shunted back to getting the energy required to run the economy if the economy is to remain the same size.

Once these necessities are taken care of then what is left is considered the *discretionary* output of the economy. This can be either discretionary consumption (a vacation or a fancier meal, car or house than needed, represented by the upper right pointing arrow in the diagrams) or discretionary investment (i.e. building a new tourist destination in Florida or the Caribbean, represented as the lowest of the top arrows feeding back into the economy. During the last 100 years the enormous wealth generated by the United States economy has meant that we have had an enormous amount of discretionary income.

It turns out that the needed information to construct the above division of the economy is reasonably easy to come by for the U.S. economy, at least if we are willing to make a few major assumptions and accept a fairly large margin of error. Inflation-corrected GDP, i.e. the size of the output of the economy, is published routinely by the U.S. Department of Commerce.

The total investments for maintenance in the U.S. economy are available as "Deprecation of Fixed Capital", U.S. Department of Commerce, various years). The minimum needed for food, shelter and clothing is available as "Personal Consumption Expenditures" (or the minimum of that required to be above poverty) which we selected from the U.S. Department of Commerce for various years). When these three requirements for maintaining the economy: investments and payments for energy, maintenance of infrastructure and maintenance of people are subtracted from the total GDP then what is left is discretionary income.

We simulated two basic data streams: the U.S. economy from 1949 to 2005 (representing the growth prior to the "oil crises" of 1973 and 1979, the impact of the oil crisis and the recovery from that, which had occurred by the mid-1990s. Then we projected this data stream into the future by extrapolating the data used prior to 2005 along with the assumption that the EROI for society declined from an average of roughly 20:1 in 2005 to 5:1 in 2040.

Results of Simulation

The results of our simulation suggest that discretionary income, including both discretionary investments and discretionary consumption, will move from the present 50 or so percent in 2005 to about 10 percent by 2050, or whenever (or if) the composite EROI of all of our fuels reaches about 5:1.

Discussion

Individual businesses would be affected by having their fuel costs increase and, for many, a reduction in demand for their products. This simultaneous inflation and recession happened in the 1970s and is projected to happen into the future as EROI for primary fuels declines.

The "stagflation" that occurred in the 1970s was not supposed to happen according to an economic theory called the Phillips curve. But an energy-based explanation is easy (e.g. Hall 1992). As more money was diverted to getting the energy necessary to run the rest of the economy disposable income, and hence demand for many non-essential goods and services, declined, leading to economic stagnation.

Meanwhile the increased cost for energy led to inflation, as there was no additional production that occurred from this greater expenditure. Although unemployment increased overall during the 1970s it was not as much as demand decreased, as labor at the margin became relatively useful compared to increasingly-expensive energy.

Individual sectors might be much more impacted as happened in 2005, for example, with many Louisiana petrochemical companies that were forced to close or move overseas when the price of natural gas increased. On the other hand alternate energy businesses, from forestry operations and woodcutting to solar devices, might do very well.

When the price of oil increases it does not seem to be in the national or in corporate interest to invest in more energy-intensive consumption, as Ford Motor Company seems to be finding out with its large emphasis on large SUVs and pickup trucks. We are likely to have over invested already in the number of remote second homes, cruise ships, and Caribbean semi-luxury hotels, so that it may not a particularly good idea to do more of that now. This is due to the "Cancun effect" – that such hotels require the existence of large amounts of disposable income from the US middle class and cheap energy.

That disposable income may have to be shifted into the energy sector with less of an opportunity cost to the economy as a whole. Investors who understand the changing rules of the investment game are like', to do much better in the long run.

So what can the scientist say to the investor? The options are not easy. As noted above worldwide investments in seeking oil have had very low monetary returns in recent years.

Investments in many alternatives may not fare much better. Ethanol from corn projects were once financially profitable to the individual investor because they have been highly subsidized by the government, but they are probably a poor investment for the Nation.

It is not clear that this fuel makes much of an energy profit, with an EROI of 1.6 at best, and less than one for one at worst, depending upon the study used for analysis. Biodiesel may have an EROI of about three to one. Is that a good investment? Clearly not relative to remaining petroleum, but some day as petroleum EROI declines it may be. However real fuels must have EROIs of 5 or 10 or more returned on one invested to not be subsidized by petroleum or coal in many ways, such as the construction of the vehicles and roads that use them.

Other biomass, such as wood, can have good EROIs when used as solid fuel but face real difficulties when converted to liquid fuels, and the technology is barely developed. The scale of the problem can be seen by the fact that we presently use more fossil energy in the US than is fixed by all green plant production, including all of our croplands and all of our forests (Pimentel, D.

Personal communication). Biomass fuels may make more sense in nations where biomass is very plentiful and, more importantly, where present use of petroleum is much less than in the US. Alternatively one might argue that if we could bring the use of liquid fuels in the United States down to, say, 20 percent of the present then liquid fuels from biomass could fill in a substantial portion of that demand.

Nevertheless we should remember that historically we in the U.S. have used energy to produce food and fibre, not the converse, because we have valued food and fibre more highly. Is this about to change?

Energy return on investment from coal is presently quite favourable compared to alternatives (ranging from perhaps 50:1 to 100:1), but the environmental costs are probably unacceptable as the case for global warming and other pollutants from coal burning becomes increasingly clear.

Injecting carbon dioxide into some underground reservoir seems unfeasible for all the coal plants we might build, but it is being pushed hard by many who promote coal. Nuclear has a debateable moderate energy return on investment (5–15:1, some unpublished studies say more), but newer analyses need to be made.

Nuclear has a relatively small impact on the atmosphere, but there are large problems with public acceptance and perhaps safety in our increasingly difficult political world. Windmills have an EROI of 15–20 return on one invested, but this does not include the energy cost of back up or electricity "storage" for periods when the wind is not blowing.

They make sense if they can be associated with nearby hydroelectric dams that can store water when the wind is blowing and release water when it is not, but the intermittent release of water can cause environmental problems. Photovoltaics are expensive in dollars and presumably energy relative to their return, but the technology is improving. One should not be confused by all claims for efficiency improvements because many require very expensive "rare-earth" doping materials, and some may become prohibitively expensive if their use expands greatly (e.g. Andersson et al. 1998). According to one savvy contractor the efficiency in energy returned per square foot of collector has been increasing, but the energy returned per dollar invested has been constant as the price of the high end units has increased (Blair May,WaldoboroMaine, personal communication). Additionally while photovoltaics have caught the public's eye the return on dollar investment is about double for hot water installations. Windmills, photovoltaics and some other forms of solar do seem to be a good choice if we are to protect the environment, but the investment costs up front will be enormous compared to fossil fuels.

Energy and money are not the only critical aspects of development of energy alternatives. Recent work by Hirsch et al. (2005) has focused on the investments in time that might be needed to generate some kind of replacement for oil, should that be possible and peak oil occur. They examined what they thought might be the leading alternatives to provide the US with liquid fuel or lower liquid fuel use alternatives, including tar sands, oil shales, deep water petroleum, biodiesel, high MPG automobiles and trucks and so on.

They assumed that these technologies would work (a bold assumption) and that an amount of investment capital equal to "many Manhattan projects" (the enormous project that built the first atomic bomb) would be available. They found that the critical resource was time – once we decided that we needed to make up for the decline in oil availability these projects would need to be started one or preferably two decades in advance of the peak for there not to be severe dislocations to the US economy.

Given our current petroleum dependence, the rather unattractive aspects of many of the available alternatives, and the long lead time required to change our energy strategy the investment options are not obvious. This, we believe, may be the most important issue facing the United States at this time: where should we invest our remaining high quality petroleum (and coal) with an eye toward insuring that we can meet the energy needs of the future. We do not believe that markets can solve this problem alone or perhaps at all. Research money for good energy analysis unconnected to this or that "solution" simply are not available. Fortunately some private individuals are stepping into the void as per our acknowledgements.

Human history has been about the progressive development and use of ever higher quality fuels, from human muscle power to draft animals to water power to coal to petroleum. Nuclear at one time seemed to be a continuation of that trend, but that is a hard argument to make today.

Perhaps our major question is whether petroleum represents but a step in this continuing process of higher quality fuel sources or rather is the highest quality fuel we will ever have on a large scale. There are many other possible candidates for the next main fuel, but few are both quantitatively and qualitatively attractive. In our view we cannot leave these decisions up to the market if we are to solve our future climate or peak oil problems.

One possible way to look at the problem, probably not a very popular one with investors, is to pass legislation that would limit energy investments to only "carbon-neutral" ones, remove subsidies from low EROI fuels such as corn-based ethanol, and then perhaps allow the market to sort out from those possibilities that remain.

A difficult decision would be whether we should subsidize certain fuels. At the moment alcohol from corn is subsidized three times: in the natural gas for fertilizers, the corn itself through the Department of Agriculture's 100 or so billion dollar general program of farm subsidies, and the additional 50 cents per liter subsidy for the alcohol itself.

It seems pretty clear that the corn-based alcohol would not make it economically without the subsidy as it has only a marginal (if that) energy return. Are we simply subsidizing the depletion of oil, natural gas (and soil) to generate an approximately equal amount of energy in the alcohol? Wind energy appears to have about an 18:1 EROI, enough to make it a reasonable candidate, although there are some issues relative to backup technologies for when the wind is not blowing. So should wind be subsidized, or allowed to compete with other zero emission energy sources? A question might be the degree to which the eventual market price would be determined by, or at least be consistent with, the EROI, as all the energy inputs (including that to support labor's paychecks) must be part of the costs. Otherwise that energy is being subsidized by the dominant fuels used by society.

CONCLUSION

It seems obvious to us that the U.S. economy is very vulnerable to a decreasing EROI for its principle fuels, whether that comes from an increase in expenditures overseas if and as the price of imported oil increases more rapidly than that of the things that we trade for it, or as domestic oil and gas reserves are exhausted and new reservoirs become increasingly difficult to find, or as we turn to lower EROI alternatives such as biodiesel and or photovoltaics. We do not know exactly what all this means, but our straightforward model suggests that a principal effect will be a decline in discretionary income and a greater investment requirement for getting energy, with all the economic impacts that entails. Since more fuel will be required to run the same amount of economic activity the potential for environmental impacts increasing is very strong. On the other hand protecting the environment, which we support strongly, may mean turning away from some higher EROI fuels to some lower ones. We think all of these issues are very important yet are hardly discussed in our society or even in economic or scientific circles.

6

Wind Power: Benefits and Limitations

Wind turbines have a potential benefit insofar as they have a power density that matches coal, at least according to one measure. Set against this is the uncontrollable nature of their output.

This means that without a suitable method of storing output, wind power can satisfy only about 10% of total energy demand. This limit applies to all uncontrollables collectively, with the slight exception that in places using a lot of air conditioning, photovoltaics could be used to help satisfy peak electrical demands.

The basic problem of uncontrollables would resolve if a suitable method of storing electricity could be found. The severe limitations of hydro, hydrogen storage, and vanadium batteries are explored. A storage system that would be both efficient and significant in size, at least in the USA, is Compress Air Energy Storage (CAES), but more experience of this is needed before it can be properly assessed.

Assessment becomes even more difficult when looking ahead to the time when all fossil fuels are scarce, because at present there appears to be no satisfactory solution to the 'liquid' fuel problem, yet the process of manufacturing, installing, and maintaining wind turbines and the associated transmission lines would be very difficult without the help of liquid fossil fuels.

In the USA, any likely gain from the use of wind power is likely to be overtaken by the present population growth of at least 1.4% a year.

INTRODUCTION

The power density1 that can be achieved using each specific renewable energy source is an important measure of the usefulness of that energy source. To see wind turbines in perspective, it is helpful to look first at a variety of energy sources.

The power density that is likely to be achieved when coal is used to produce electricity has been estimated at 315 kW(e)/ha.2 Note that the power density is there given in terms of the electrical output. Since the efficiency of producing electricity from coal is about 30%, it can be deduced that, in terms of the coal that produces the electricity, its power density is about 315/0.30 = **1050** kW/ha. The normal route is of course first to calculate the power density of coal itself, but that is incidental.

After establishing the output of electricity from wind turbines, as will be done later, it will be appropriate to discuss whether emphasis should be placed on the power density in terms of just the electrical power produced from the wind turbines or whether, as is often done, that output should be uprated to take account of the fossil fuel required to produce it. For the present, note only that as the output of wind turbines is electricity, the first step will be to measure the power density in terms of the electrical output, i.e. power density measured as kilowatts of electricity, **kW(e)**, rather than **kW** of fossil fuel equivalent.

Before proceeding further into the study of wind power, it will be relevant to look briefly also at the power density of liquid fuels produced from biomass. There are various categories of power density which can be assessed, all of them useful in their own way. The one that is least controversial is to measure the output per hectare of, for example, ethanol, subtracting from it only the amount of energy input that needs to be in liquid form, e.g. as gasoline, diesel or ethanol. That gives the 'useful' ethanol per hectare. In such an assessment, the power density of ethanol from corn (maize) is about 1.9 kW/ha (OPTJ 3/1).4 Incidentally ethanol from sugarcane, when assessed on this same basis, typically achieves a power density of 2.9 kW/ha, but soil erosion problems are worse with sugarcane than with corn, and the land that is suitable for growing sugarcane is more restricted. Considered against the power density of oil, which is considerably higher than the 1050 kW/ha mentioned for coal, it is clear that these ethanol power densities are very small indeed. For example, in the same paper, OPTJ 3/1, it is calculated that if all the U.S. corn crop were to be used to produce ethanol, it could serve to replace only 6% of the fuel used in the USA for transport.

Another type of power density that can be assessed is by adding to the ethanol output the calorific value of the by-products (e.g. dry distillers' grains that can be fed to cattle), and from that subtracting not only the liquid input but also the nonliquid inputs, e.g. the heat needed for distillation (which constitute about 85% of the inputs).

The resultant 'net energy capture' would be a revealing figure if its value could be agreed, but there are huge areas of uncertainty, particularly because we need to know (a) how much of the by-product is actually going to find a use and should therefore be counted as an output; (b) how much of the total crop can be utilized without causing loss of soil quality. For example, in the case of corn total yield is about 15,000 kg/ha (dry), with about half of this being grain and the other half being stover (Pimentel and Pimentel 1996, p. 36). Growing corn is prone to cause soil erosion. All the stover should be either left on or returned to the ground to diminish erosion and return nutrients. Sugarcane is worse than corn at causing soil erosion (Pimentel 1993), so a very significant proportion of the bagasse should be returned to the soil rather than using most of it to produce the heat needed for ethanol distillation (as tends to be done in practice).

All energy balance calculations are crude at best due to such factors, and the 'energy balance' of producing ethanol from corn can be assessed as either positive or negative depending on matters of fine judgement. However, let us be clear about what an approximate zero energy balance means. It means that producing ethanol from biomass is not an energy transformation that produces *useful energy;* it is merely a way of using other forms of available energy to produce energy in a liquid form. The conclusion is twofold: that power density figures need to be hedged about with precise understanding of what is being assessed, and that producing significant quantities of liquid fuels from renewable sources is a difficult problem.

THE POWER DENSITY OF ELECTRICITY FROM WIND TURBINES

In an ideal situation, where the wind always blows from the same direction, and where docile citizens do not mind where the wind turbines are placed, the turbines could be placed fairly close together. But in practice there are few sites where engineers believe that the wind can be trusted to always come from the same direction. Moreover there are often practical restrictions about where the wind turbines can be placed. Due to these factors, the actual placing of wind turbines is such that about 25 ha needs to be 'protected' from interference by other wind turbines for each megawatt (MW) of wind turbine *capacity* (Hayden 2004, pp. 145–149). Note first that this 25 ha/MW is independent of the rated capacity of the wind turbine (e.g. two turbines of 1MW capacity would require 50 ha and so would one 2MW turbine), and secondly that the 25 ha/MW refers to the *rated capacity* of the wind turbines not their actual output.

The actual output of a wind turbine, or group of wind turbines, is determined by the capacity factor (also called load factor) that they achieve. In northern Europe (Sweden, Denmark, Germany, the Netherlands) the mean capacity factor achieved over two years was 22% (OPTJ 3/1, p. 4), in the UK for the years 2000–2004 capacity factors achieved were 28%, 26%, 30%, 24%, 27% for an average of 27%,6 and for the USA for the years 2000–2004 capacity factors were respectively 27%, 20%, 27%, 21%, 27% for an average of 24%.7 Nevertheless taller wind turbines may produce some improvement, so let us use 30% as a benchmark for the USA. This means that the protected area is 25/0.30 = **83** ha per MW *of output*, which gives a power density of 1000 [kW(e)]/83 = **12** kW(e)/ha.

That power density gives an easy way to calculate how much land area would be needed to provide a certain amount of electrical output; e.g., to produce the mean power output of a 1000MW power station, which delivers over the year say a mean 800MW, the area needed would be 800,000 [kW]/12 = **66,700** ha, or 667 km^2, or 26 km by 26 km (16 miles by 16 miles). That is a substantial area, the ramifications of which will be considered later, after some other measures of power density have been considered. Also of considerable relevance is the amount of land that the wind turbines are actually taking up, that is the land taken up by the concrete bases of the turbines and transmission lines, and to provide access roads (obviously this is mainly of concern when the land that is being used is ecologically productive). This has been put at 2–5% of the protected area. Taking a central value of 3.5%, puts the power density of wind turbines — in these terms, when sited on ecologically productive land — at 12/0.035 = **343** kW(e)/ha. That is to say, it is similar to the power density of electricity from coal. It now becomes obvious why wind turbines are in a different ball park from biomass; that holds true whether the biomass is used to produce ethanol or merely used for its heat value. To touch on the latter briefly, it may well be possible to achieve, at suitable locations, without too many inputs, an annual yield of 10 dry tonnes per hectare using woody short-rotation crop. That would achieve a *gross* power density of about 6 kW/ha. Note that both the wind power density figures being discussed, as well as the 6 kW/ha biomass figure, should really be qualified with the adjective 'gross', because no allowance has been made for inputs. However the difference between 6 kW/ha and 343 kW(e)/ha is so great that it is not necessary to determine to what extent inputs bring the *net* power densities closer together.

PRODUCING THE OUTPUT OF A POWER STATION FROM WIND POWER

Returning to the calculation which showed that to replace a 1000MWpower station by wind farms would require 667 km2 of protected space, a small point to address first is the choice of 800MW as the mean output. That may be challenged on the basis that power stations generally operate below an 80% load factor. The point though is that many power stations operate below capacity simply because they are *controllable*, which allows their output to be adjusted to suit demand. Clearly wind power cannot be used in that way. Instead it is used in conjunction with a controllable power source. The two operate together, 'in harness', to provide a baseload. Plant operated in that way, that is just to provide a baseload, e.g. nuclear plant, can certainly achieve an 80% load factors. Hayden (2004, p. 246) shows that 7 out of 22 countries operate their nuclear plant at above 80% load factor. The practical problems of needing such large areas over which to spread the wind turbines is particularly acute in places with high population densities like Europe. But difficulties are encountered in practice in the USA too, due to such things as objections to destroying scenic vistas by putting wind turbines along prominent ridges. Moreover there are other problems in the wide spacing when taking a longer view.

The mean 800MW of output, with a 30% load factor, would require a capacity of 800/0.3 = **2670MW**, which might be supplied by 888 wind turbines of 3MW capacity, for example. The task of installing those, with their access roads, and then connecting them together over an area of 667 km2, may not seem too daunting to an engineer in the present day, but that is only because fossil fuel oil is available. When liquid fossil fuels become scarce, and a renewable liquid substitute has to be used, most probably one with something like the low power density we considered for ethanol from corn or sugarcane, the challenge would become enormous. *In planning* for a fossil free future, it is necessary to continually bear in mind that many things which are easy today because of the availability of suitable fossil fuels, particularly *oil, will not be easy in the future.* Whether such tasks as installing and maintaining wind turbines and transmission lines will be possible in the virtual absence of oil must at present be a matter of judgement.

THE PROBLEM OF ASSESSING *NET* ENERGY WITH RESPECT TO WIND TURBINES

Net energy is simply the energy left over as useful energy once all the inputs have been subtracted. While that is a simple concept, there are practical problems which it is worth dwelling on. The wind industry would most likely respond to the previous paragraph by saying that the 'energy payback' — which is the time it takes to produce enough energy from the wind turbines to produce the same amount of energy as the inputs that are needed for their construction from raw materials and subsequent maintenance — has already been assessed for wind turbines, and it has been put as low as six months, so there must be something misleading in the emphasis being placed on the extent of the inputs needed as per the previous paragraph.

The trouble with such energy payback assessments is that they take only partial account of the different *types* of input and sometimes they do so in a misleading way. For example, in assessing the energy value of the electrical output of wind turbines, that output is valued as the amount of fossil fuel that would be needed to produce it. Since the efficiency of generation of electricity is about 0.33, that means that the electrical output can be valued at 1/0.33 = **3** times its energy value as electricity. There is some validity in that when electrical energy is so useful to us that society is prepared to suffer the unavoidable loss of energy that occurs in producing it from fossil fuels. However, looking towards a fossil-fuel-free society, the situation is entirely different. We have already noted that the power density of a renewable liquid fuel is below 2 kW/ha, and that of biomass when used merely as heat is around 6 kW/ha, so it would be sound sense to use the high power density of wind turbine output (12 kW(e)/ha or 343 kW(e)/ha depending on the perspective) to replace both heat and *if it is possible* use the electrical output to produce 'liquid' fuels.

Thus far from electricity being at a premium value, it is either at no premium, because it is used to replace the heat needed for such industrial processes as glass making, or at a substantial discount in value, because of the large losses that would occur in trying to produce a useful 'liquid' fuel from it, e.g. compressed hydrogen. The extent to which that is viable is a relevant question to be addressed later. What has become apparent is that wind turbines have a far higher energy density than biomass, on one measure even rivalling that of coal, so the next consideration is to what extent it is advisable to integrate the input from wind turbines into the electrical system just to save fossil fuel now, while we still have the oil to carry out the construction, installation and maintenance processes associated with wind turbines without too much difficulty. That leads on to consideration of the problems of dealing with the *uncontrollable* nature of the output from wind turbines.

The Implications of the Uncontrollable Nature of the Output from Wind Turbines

To fully understand the problem that uncontrollable inputs of electrical power introduce, perhaps it is best to consider an extreme situation, just to see what effect that would have. Such an extreme is entirely unrealistic, but it will serve to clarify the general principle.

So take, for an imaginary example, a situation in which a widespread group of wind turbines do sometimes produce their full rated power. To be slightly more precise, let us say that the wind turbines are as widely spread as the E.ON Netz network in Germany which covers a distance of 800 km. The assumption of an output of full rated power means, of course, that it is thereby assumed that *at times* the wind blows sufficiently hard to allow every single turbine to produce at its rated power. That is fanciful, but let us now make an even more fanciful assumption that at other times over the course of the year the wind is so desultory that these wind turbines produce only 5% of their rated power. It is immediately obvious that these turbines would be useless for following variations in consumer demand. For that purpose, demand-following plant would have to be used. The only use that could be made of the input from the wind turbines would be to run them 'in harness' with controllable plant which would produce the remaining 95% of the rated power of the wind turbines. Working in harness, the wind turbines and the controllable plant together could produce a baseload equal to the rated power of the wind turbines. In such a clear-cut and extreme situation that is obvious to common sense. Although the actual situation is more complicated, a similar principle applies in reality.

As already suggested, a suitable benchmark for the capacity factor (also called load factor) of wind turbines is 30%. The 'peak infeed' from wind turbines is defined as the highest output they will reach as a proportion of their rated capacity. Statistics on this parameter are hard to come by except from the distributor E.ON Netz whose network, as mentioned, extends over 800 km. The documentation of their experience from operating wind turbines is superb.9 From their experience over two years, it seems that peak infeed from their widely spread turbines is about 80% of the rated capacity of the wind turbines. Following the same *principle* as in the previous imaginary example, it can be deduced that in these circumstances wind could provide 30/80 = 38% of the baseload block of electricity, with controllable plant filling in the remaining 62%. A recent modelling study for the UK,[10] based on taller wind turbines located at all the windiest spots spread over the entire UK, showed that during the month of January, in the twelve years studied, the average peak infeed was 98%, and in one year it was 100%. The study's estimate of capacity factor was 35.5%. Note that the all important ratio, in these more windy conditions than Germany, remains much the same, at 35.5/98 = 36%.

That is not to say that the wind can satisfy 38% of *total* electrical demand, because, as observed, wind and the plant operating in harness with it can only produce a *baseload*. If there is no nuclear plant operating which needs to be allowed to operate without restrictions to produce a baseload, then wind turbines and the plant operating in harness with them can be set the task of providing a baseload up to the level of low demand. Low demand is about 60% of mean demand. Thus wind output can satisfy 38% of 60% which is 23% of electrical demand, provided that there is no other plant (e.g. current-design, inflexible nuclear plant) that is already fulfilling part of the baseload supply. 23% of electrical demand is only about 10% of total energy demand,[11] but 10% would appear to be worth pursuing provided that it does not too much interfere with the rest of the electrical system. That is what needs to be considered next.

The Problems of Operating in Harness with Wind Turbines

The effect of introducing wind into an electrical system cannot be judged on the electrical input from wind alone. As we have seen, the task has to be shared: about 38% taken by wind and 62% by a controllable power source. When wind becomes a significant part of the whole, it degrades the efficiency of the rest of the system, not only because of the need to keep plant running to cope with sudden wind changes, but more importantly because of the need to be able to start and stop plant on a frequent basis. Plant designed to do that operates considerably less efficiently than plant optimized to run at constant load. No one knows just how much less efficiently plant actually operates when it has to run in harness with wind turbines, however the effect is not small. In the extreme case of an all-natural-gas system, it can be shown that the loss of efficiency of the plant operating 'in harness' outweighs the benefits of the wind input (OPTJ 5/2, pp. 8–17). In conclusion, while maximum integration of wind turbines may appear capable of saving 10% of fossil fuel use, the actual figure will be lower than this because of:

(a) the additional energy needed to construct and maintain the turbines, and

(b) the degraded load factor and efficiency of the plant when it operates in harness with the wind turbines.

Also to be borne in mind is that even if the full 10% could be saved, this would rapidly be eaten up by population growth in the USA; a point we will now turn to. Electrical production in the USA in 2005 was about 3.8 billion MWh. 23% of that is 0.87 billion MWh or an annual mean power output 99,000MW. Thus 99,000/800 = **124** wind turbine farms, each producing a mean 800MW, would be needed to provide the electricity. They would cover a total area of 124×667 km2 = 83,000 km2. It is hard to imagine such a task being accomplished under a decade. Before the decade was out, the 10% of energy demand saved by introduction of the wind turbines would be overtaken by the increase in energy demand due to population growth, as can easily be seen. During the final three decades of the last century, the rate of population growth in the U.S. was 1.06% per year. Even at that growth rate (and it is now higher), by the end of the decade of frantic wind turbine installation, population would have grown by 11%, increasing total energy demand by 11%, and thus outstripping the 10% of energy saved by the newly installed wind turbines. The extent of public opposition can be judged by the fact that so far wind contributes only 0.4% to electrical production in the USA, and that has already caused vociferous complaint. It should be mentioned, too, that the 1.06% per year is an understatement, as it has recently been shown that by the time all the illegal aliens are accounted for, the present rate of population growth in the U.S. is probably in the range of 1.4–1.7% (Abernethy 2006).

ALTERNATIVES TOWIND POWER

What is often not appreciated is that there is a limit to the contribution from uncontrollable power sources in an electrical supply system. It has been shown that wind turbines can only contribute about 23% of total electricity. A double share could not be achieved by allowing another uncontrollable, say wave power, to also produce 23%. The wave and wind power generators would sometimes produce their maximum output at the same time and thus overwhelm the electrical system. It is therefore necessary to choose only the best form of uncontrollable available at a given time. It should be mentioned that photovoltaics may be an exception, at least in a country that makes heavy use of air conditioning. This is because although peak demand tends to be later than midday, and it is likely to become even later as better insulated houses are built, nevertheless demand at midday will be well above the minimum demand, so to some extent photovoltaics could, cost permitting, reduce fuel use without interfering with other uncontrollables (which are limited to operating below minimum demand). With all other uncontrollables the output correlates poorly with demand; that is true even if the time of output is predictable as it is with tidal flow energy. Thus without storage, it becomes necessary to choose, and go for the best type, provided of course there is sufficient potential output available from that type.

It is clear that wind power has many problems. These stem chiefly from the capacity factor being small in relation to peak infeed, and partly because it is hard to forecast the output from wind to within a few hours, which is desirable for the efficient operation of the plant that has to operate in harness with it. Installation of wind turbines is termed by some as an industrialization of the landscape and, while it is impossible to put a value on the loss of quality of life that would occur for many people thus afflicted, one should not lose sight of that aspect. A further adverse effect of wind turbines is a significant slaughter of birds and bats.12 Together all these factors suggest that every endeavor should be made to research wave power. Wave power would certainly be more predictable and less prone to sudden change, and it might offer a better ratio between its capacity factor and peak infeed, thus enabling it to take a larger share of the total demand for electricity than wind ever could. Whether it could be made economically viable is of course another matter.

THE PROBLEMS OF STORAGE

The foregoing has not presented a cheerful prospectus for uncontrollables. What everyone hopes is that the problem of uncontrollables will be overcome by finding a way of storing the energy. Storage would solve the problem of not only wind but all uncontrollables, so it deserves detailed consideration. **Hydro.** The most useful way to store electricity is in the form of water in a reservoir — using 'pumped storage'. That can be excellent for small amounts of electricity, but calculation soon shows that the capacity available is small compared to the requirements of large populations, especially when it is borne in mind that to produce a steady supply of electricity from wind turbines, only 38% of the block of electricity (according to the above calculation) could be delivered directly, while the remaining 62% would need to be stored first.

Some insight into the problem is gained by looking at the power density of the average reservoir. Based on a random sample of 50 U.S. hydropower reservoirs, ranging in area from 482 ha to 763,000 ha, it has been calculated that the area of reservoir needed to produce 1 billion kWh/yr (a mean 114,155 kW) is 75,000 ha (Pimentel and Pimentel 1996, p. 206). Thus over the course of a year, the power density achieved by these reservoirs is 1.5 kW(e)/ ha. The low power density of water storage arises because to store the energy of 1 kWh, the amount of water which must be raised through 100m is 3.67 tonnes (3.67m3).

And allowing for an overall 75% efficiency in using electrical pumps to elevate the water and then using turbines to regenerate the electricity, 3.67/0.75 = **4.9** tonnes of water must be raised through 100m in order to store 1 kWh(e).

To store one week's output from a 1000MW plant, running at 80% capacity, would require *660 million tonnes* of water to be raised through 100 m. To put it another way, the area of this substantially elevated reservoir would need to be 66 km2, or 8 km by 8 km (5 miles by 5 miles), and it would need to tolerate the water level being raised by 10 m. Suitable reservoirs of this kind are hard to come by, quite apart from the extra problem of needing a lower reservoir to hold the water waiting to be pumped back up.

Hydrogen. It is frequently proposed that electrical energy could be stored as hydrogen. There are many problems with that, the first being efficiency of transformation. Hydrogen production by electrolysis is around 70% efficient. About the best efficiency to be expected from fuel cells, including the need to invert their direct current output to AC, is 60%. That makes an overall efficiency of 0.70 × 0.60 = 42%.

So to deliver 1 kWh of stored electricity 2.4 kWh would have to generated from the wind turbines, and that is without allowing for further losses in compression which is likely to be necessary for realistic storage of a gas which has an energy density approximately a quarter that of methane (natural gas).13 For an extended treatment of the problems, see *Hydrogen and Intermittent Energy Sources*, OPTJ 4/1 (pp. 26–29).

Vanadium batteries. Batteries are a possibility, particularly those which store the electrical energy in the form of a liquid in tanks which are separate from the 'engine', for this would appear to offer unlimited expansion using many tanks. A vanadium battery of this kind has been developed, but Trainer (1995, p. 1015) points out various limits, one being that the US Bureau of Mines states that demonstrated world recoverable resources of vanadium total about 69 billion kg.14 So shortage of vanadium might set an ultimate limit to producing vanadium batteries; but before considering that, let us look at problems concerning the amount of hardware that is needed.

Considerable work has gone into development of vanadium batteries since Trainer's paper. In the 13 January 2007 issue of *New Scientist* there was a three page report on the type of batteries which are being installed by an Australian firm named in the article as Pinnacle VRB. The title of the article, by science journalist Tim Thwaites, was *A Bank for the wind: at last we can store vast amounts of energy and use it when we need it.* While little trust should be placed in the titles of articles in *New Scientist* or other popular science magazines, that does suggest the need for a closer look at the potential of vanadium batteries. After describing how some of the problems of vanadium batteries had been overcome, the article had this to say: After more than a decade of development, Skyllas-Kazacos's technology was licensed to a Melbourne-based company called Pinnacle VRB, which installed the vanadium flow battery on King Island.With 70,000 l of vanadium sulphate solution stored in large metal tanks, the battery can deliver 400kW for 2 h at a stretch.

Those figures indicate that 87 liters of vanadium sulfate are required to store 1 kWh. A source in the firm has confirmed to me that the figure is approximately correct, and that 70 liters per kWh are used at the planning stage. That is a very low power density. As 1 liter of gasoline contains about 9.3 kWh, it would take 650 liters of vanadium sulfate to store the energy contained in a liter of gasoline. Even in stationary situations, such a low energy density seems likely to engender problems in terms of net energy, because the inputs required to provide and maintain the hardware may become so large as to use most of the output. To consider the overall problem we need to have an idea of how much storage is likely to be required.

Since wind is fairly low for some months, there needs to be storage to cover the low wind months. There are no figures available for the USA, butWindstats provide good month by month data for Denmark, Germany, Netherlands, and Sweden. During the months of May thru September in the two years 1998/1999 and 1999/2000, the shortfall in terms of the missing kWh (that is missing on the supposition that delivery needs to be constant each month) through those months, expressed as a fraction of the total year's delivery, was as follows for the two years: Denmark, 14.0%, 9.2%, Germany 13.8%, 14.4%, Netherlands 13.6%, 15.8%, Sweden 13.6%, 15.8%. Considering that just two years of observation are unlikely to have covered the most extreme situation, we may need something more than the worst result of 15.8%, but there is no need for too much accuracy so let us settle for storing 16% of the total annual output to cover the low wind months.15 Storage efficiency also needs accounting for. By time the AC output of wind turbines has been changed to DC, and the DC output from the VRBs has been returned to AC, the overall efficiency is probably about 70%, but let us use 75%, resulting in a need to send for storage 16/0.75 = 21% of the annual output of the wind turbines.

Before proceeding with the calculation, there is a possible objection that should be addressed. It may be thought that it is not *really* necessary to be able to store enough energy.Would itmatter if for a couple of weeks every two years wind turbine storage was exhausted and thus made peak demands worse by failing to contribute when needed? The answer is that it would matter, because available fossil fuel capacity would have to be kept available just to satisfy those rare occasions when the problem of peak demand were exacerbated by shortfall of wind energy (because it could not maintain its prescribed baseload).

In terms of a plant that delivers a mean 800MW, the amount to store, 21% of that, amounts to 1470 × 106 kWh. At 70 liters per kWh that would require 103 million cubic meters of electrolyte. Using large storage tanks, say 20m in height and diameter (about 6300m3 capacity), 16,300 such tanks would be needed. The surface area of one cylindrical tank would be 1885m2. The total area would be 30.7 million m2.

Assuming that steel with an average of 10mm thickness is used, that is 307, 000m3 of steel, or about 2.46 Mt or 2640 million kg. The embodied energy in steel is about 21 kWh/kg (Pimentel and Pimentel 1996, p. 206), so the energy embodied in the steel containers alone would be at least 51×109 kWh.16 The annual output of a 1000MW plant running at 80% capacity would be 7 × 109 kWh, so the steel for delivery of 16% of output after storage alone would cost over seven years of output, without including other construction energy costs associated with storage.

In addition to storage requirements, there would be the 'engine' component. To produce the mean 800MW from wind turbines, with a 30% capacity factor, 800/0.30 = **2667**MWof rated capacity would be required.With an 80% peak infeed this would sometimes produce 2667 × 0.80 = **2130**MW.

However 800MW of this would be used directly (to maintain the base load of 800MW, and only the remaining 1330MW would be an 'overflow' and need to be sent to charge the battery. A 1.5MW battery system currently being installed requires an 'engine' of about 45 tonnes (50m3). On that basis, to provide 1330MW of battery power would require 40,000 tonnes of material for the 'engine' component. The high dollar cost of the 'engine' component indicates a likely high embodied energy cost.

There are certainly advantages in vanadium batteries. For instance the electrolyte never 'wears out', having a virtually infinite life. But the above figures suggest that until the energy balance calculations have been done, it is idle to claim *'at last we can store vast amounts of energy and use it when we need it'*.

The energy inputs need to cover installing and maintaining the wind turbines, transmission lines, plus tanks for electrolyte storage, plus the 'engine' component of the battery and inverters to produce AC current from the DC output. But it is just possible that the outcome on energy balance will look acceptable, so let us turn back to the question of availability of vanadium.

Earlier it was noted that wind turbines might contribute 23% of mean demand, which in relation to the USA could be expressed as an annual mean power output of 99,000MW. We have also noted the need to store 21% of that output in order to produce a steady baseload through the less windy months. Thus a mean 20,800MW = 182 billion kWh would need to be stored. At 0.39 kg of vanadium per kWh,18 that would require 71 billion kg of vanadium. Yet we noted above that the US Bureau of Mines states that demonstrated world recoverable resources of vanadium total about 69 billion kg. Cost would also be a likely barrier.

Clearly even if the energy balance is better than it appears *prima facie*, although vanadium batteries might assist the USA in delivering from store 23% × 0.16 = 3.2% of its annual electrical consumption, they cannot provide a worldwide solution, and not much of a solution for the USA, for integration of this storage plant would merely enable the 23% of total electricity which is to be produced from wind to be stabilized at 30% of the rated capacity of the wind turbines (thus avoiding the need to use fossil fuel plant to work in harness).

While there is no theoretical bar to installing more wind turbines and vanadium batteries to cover more of U.S. electrical supply than 23%, it is clear that the availability of vanadium means that there is little scope for that, even if the cost were to be bearable. It should be noted that a storage requirement of 21% of the output of the wind turbines serves only to sustain output through any one year. There is another problem.

The U.S. capacity factors in 2001 and 2003, were 20% and 21% respectively. Were the aim to be to provide a reliable output from wind (thus obviating the need to keep fossil fuel back-up for rare occasions), so as to be able to guarantee to produce in every year the 27% capacity factors of 2000, 2002 and 2004, it would be necessary to store 1–(20/27) = 26% of the wind turbine's best annual output, i.e. that achieved with a 27% capacity factor. This would be needed in order to top up the 20% load factor of 2001 to 27%.

Moreover to deliver that 26% would require 26/0.75 = 35% to be sent to storage. This 35% is not instead of the 21% calculated previously but in addition to it. Again it will doubtless be asked whether that is really necessary.

Again the answer is that it is not, but to the extent that the storage is not available, a controllable output is needed which can be brought into action during the years in which the wind fails to come up to scratch. The difficulties in making use of an uncontrollable output are very great.

There are other possible batteries, such as nickel-cadmium, sodium-sulfur, and sodium-nickel-chloride, but sufficient data are not available to assess their potential. The above look at vanadium batteries has been concerned with their effectiveness in solving the overall problem of wind uncontrollability. In that respect, the limitations have been made evident, but perhaps it should be mentioned that there are some limited uses for them provided the cost is tolerable.

For instance, Japan has such gusty winds that it is a problem integrating the output from wind turbines. A vanadium battery can be used to damp the wilder excursions. Also it has been suggested that vanadium batteries could take all the output of wind and then sell the output at a much higher price for satisfying peak demands. The principle is sound, but there is insufficient data to determine whether this is going to prove economically viable.

CAES. Another method of storing electrical energy is compressed air energy storage, CAES, in which air is compressed and stored underground. The compressed air is later used to increase the output of gas turbines by about 200% (by saving the two-thirds of the energy output that would normally go into compression). However the extent of the problem arising from low energy density exceeds even that of hydropower. There are two operational CAES plants.

The plant at Huntorf, located in North Germany, was commissioned in 1978 and has been in operation ever since. It is designed to hold pressures up to 100 bar although 70 bar (1015 psi) is set as the maximum permissible operational pressure.

Information available for it20 suggests that under normal storage, within the 310, 000m3 space, energy density is about 2 kWh/m3. However there are several ambiguities in the precise meaning of the data, including uncertainty about whether the quoted 300MW output for 2 h results partly from the natural gas used. Certainly the figure of 2 kWh/m3 energy density appears high in comparison to the McIntosh CAES plant of the Alabama Electric Company, commissioned in 1991.

Moreover the McIntosh plant is said to include 'several improvements over Huntorf, including a waste heat recovery system that reduces the fuel usage by about 25%'. The maximum pressure for storage is reported as being 74 bar (1070 psi), and it is stated that the 5.32 million m3 cavern can deliver power at 110MW for 26 h. That indicates an energy density of storage of only 0.54 kWh/m3.

At certain places in the world, the available storage space is vast. I have been assured by an experienced operator in the electricity industry that, in Alabama, 'we are aware that there is tight gas storage of at least 548 billion cubic feet capacity with constant 750 psi pressure from hydro aquifer support'. 548 billion cubic feet equals 15.5 billion m_3.

At the aforesaid 0.53 kWh/m3, this would make available from store 8.2 billion kWh. That is equal to the annual output of a 1000MW power station, operating at 94% capacity. But storage capacity on this scale is not readily available, and even if one is prepared to overlook the need for the turbines to run on natural gas (no commercial solution has yet been demonstrated for running the generators efficiently on compressed air alone), albeit being made more efficient by the infeed of high pressure air, CAES does not appear to offer a worldwide solution to storing electrical energy because of storage space, irrespective of how high the efficiency of the method may be (it has been put as high as 80%). It has been suggested that with the world emitting about 18 billion tonnes *excess* carbon dioxide each year by burning fossil fuels, there is a need to use most of the available storage space for storing carbon dioxide; but compressed air storage is formed in solution-mined caverns underground, basically very large 'empty' caverns.

Carbon dioxide sequestration is best made into old oil deposits for enhanced oil recovery, or into saline aquifers, which can absorb significantly higher amounts of CO2 than could be obtained from the equivalent amount of open space volume. However it should not be forgotten that the practicality of sequestration into saline aquifers remains to be established.

In summary, while fossil fuels are available, there must be doubts whether a significant amount of net energy could be produced by combining wind turbines with such limited storage capacity as could be made available to assist them. Without fossil fuels, the whole project of producing wind turbines, transmission lines, plus storage capacity and regenerators is likely to be impossible (see problems of 'liquid' fuels below).

THE PROBLEM OF 'LIQUID' FUEL IN A FOSSIL-FUEL-FREE SOCIETY

Doubt was previously cast on the possibility of constructing and maintaining large wind farms in the context of a post-fossil-fuel society. The main reason was because of the difficulty of providing fuel in a 'liquid' form.

The hope will obviously arise that the relatively high power density of the uncontrollables, including wind turbines, could be used to produce hydrogen by electrolysis. We need to ask whether that idea might be viable.

The essence of producing 'liquid' hydrogen from electricity is to produce the hydrogen from water by electrolysis and then to liquefy it, so that its energy density is sufficient to make it useful for transport. Even as a liquid, it would take 3 liters of liquid hydrogen to move a vehicle over the same distance as 1 liter of gasoline would take a similar car (OPTJ 3/2, pp. 21–27). It would take 9.1 kWh of electricity to produce liquid hydrogen with the same motive energy as 1 liter of gasoline (or 34 kWh(e) per *gallon* of gasoline). The cost of that might seem bearable, except that the output of wind turbines is erratic.

It seems unlikely that a production line could be run for producing liquid hydrogen using only the erratic input from wind turbines (which produces some, but often not much, electricity for 95% of the year). Yet the alternative of running the plant continuously would require about two thirds of the electrical energy to come from a controllable power source.

Because the efficiency of transformation in producing electricity from fossil fuels is about 33%, if for simplicity we assume for a moment that all the energy needed to produce the equivalent of 1 liter of gasoline were to come from a controllable power source, then that energy needed would amount to 9.1 [kWh(e)]/0.33 = 27 kWh.

That would be somewhat alleviated by 38% of the electricity coming directly from the wind turbines, but nevertheless such an inefficient process is unlikely to be attempted while fossil fuels are available; when fossil fuels become scarce, there would be insufficient energy available to contemplate the process. To put it another way, producing liquid hydrogen from renewable sources via a steady production process depends on getting a steady supply by supplementing uncontrollable inputs. Such supplementation could only be achieved if the problem of storage is solved. *The fact* is that at present there is no solution in sight to producing the quantities of 'liquid' fuels from renewable sources which would be required to allow present populations to live in even a very frugal version of present lifestyles.

LEARNING FROM EXPERIENCE (DENMARK)

In the above theoretical analysis, it was noted that the inefficiency introduced into the electrical system by running plant in harness with an uncontrollable power source has not been assessed. For that reason alone it is helpful to try to learn from the experience of a nation which has attempted to make maximum use of wind power, namely Denmark. Inevitably there will be other variables which distort the effect of introducing wind power into the system but some clues can be gained.

Denmark is the nation which should reveal the most about integrating wind power into its electrical system, because in 2004 the electricity produced from its wind turbines amounted to 18.5% of total electricity production.

But Denmark can only use a third of this directly, partly because of the very problem of the uncontrollable nature of the output, and partly because the greatest part of the wind turbine electricity is produced in the west of Denmark, and the west Denmark grid is separate from the east Denmark grid.21,22 This has not inhibited the development of wind power because Denmark has interconnectors to Germany, Norway and Sweden which could carry virtually the whole of west Denmark's wind output.

The latter two countries have very substantial hydropower capacity, so they can switch off their hydropower and use Denmark's electricity from wind turbines instead. The Danes can then reimport the electricity as hydropower electricity at a time that suits them (albeit at considerable expense).

Thus although Denmark does not use all its wind turbine electricity directly, wind turbines should serve to reduce its carbon emissions unless the inefficiencies of integrating wind into the system outweigh the advantages of the wind input.

Factors which might distort that assessment are that Denmark has also been trying many other things to reduce its carbon emissions through: (a) greater use of biomass; (b) extensive use of combined heat and power to provide nearly a third of west Denmark's electrical capacity; (c) a high tax on cars together with the provision of excellent public transport, (d) a high standard of insulation for its buildings.

If a substantial reduction in carbon emissions had occurred, the picture would be blurred, because any of those items might have been the reason for the reduction, but since there has not been a significant reduction, we can deduce that neither those efforts *nor the input from wind turbines* has had much effect.

To be more precise, carbon emissions per person in Denmark decreased, between 1990 and 2003, by 0.07% compared to an 8.4% decrease in the United Kingdom, which has only a 0.5% wind penetration. Admittedly the decrease in the UK was almost entirely been a result of our dash for gas—replacing coal-fired plant with powered gas generators. In 2003, Denmark's carbon dioxide emissions were 10.9 t/cap compared to the UK's 9.5 t/cap. These figures appear to prove two things.

The first is that introducing into an electrical system about 20% of the electricity from wind turbines (the most that countries are likely to be able to introduce) *may* have some effect on reducing carbon emissions, but it is hard to detect.

Secondly, it shows that when a nation tries all the things that are often proposed as politically palatable ways of reducing carbon emissions, the actual effect of reducing carbon emissions is also hard to detect.

Perhaps it should be noted that it could always be claimed that the carbon emissions in Denmark would have risen considerably more without such efforts. It could also be argued that the savings in energy use have not yet shown up due to the amount of energy being put into constructing and installing wind turbines, but such points probably do not weigh heavily, and it seems a fair conclusion that tackling only what is fairly easy in political terms does not make a significant impact on excessive carbon emissions.

MAKING REALISTIC ASSESSMENTS OF THE COST OF WIND POWER

The main thrust of this analysis has been at the fundamental level of energy. A brief comment on the potential for misleading statements about wind costs may be useful. The wind industry has for some time been saying that the cost of electricity from wind turbines is about to come down so as to be equal to the cost of electricity derived from fossil fuel. However the cost they are referring to is the total amount of money that the wind turbine operators need to be paid, *for all the kWh that they produce*, in order to bring in a satisfactory profit to the wind turbine operators. In some countries, e.g. Denmark, most wind power is 'prioritized' so that distributors have to use it. In the UK there is effectively a penalty if it is not used.

But what would be satisfactory for the wind turbine operators if all their electricity were to be bought (by whatever forms of compulsion or incentives), is very far from the real cost of wind turbine electricity.

Other costs beside those incurred by the wind turbine operator needs to be added: (1) the amortized cost to the distributor of installing, plus the cost of maintaining, the necessary additional transmission lines, and (2) the additional costs incurred when purchasing electricity from controllable sources when the controllable sources are forced to operate at lower capacity in order to make room for wind power when it is available.

The second of these is very significant. It is one thing to make a contract with the operator of a fossil fuel plant to produce a steady output, but quite another to have to make many short term contracts to top up the delivering of electricity only to the extent that wind is not able to deliver it.

CONCLUSION

Wind turbines have a potential benefit in that they have a power density that matches coal, at least according to one measure. Set against this is the uncontrollable nature of their output. Looking ahead to when fossil fuels become scarce involves consideration of the low power densities that are likely to be associated with 'liquid' energy sources. At present, it is hard to say whether building wind farms and running a grid will be possible without fossil fuels, especially because no viable renewable fuel in 'liquid' form is evident.

Concerning introducing wind turbines in order to reduce the present use of fossil fuel, while it is probable that wind turbines do save some fossil fuel, there is no evidence of this from Denmark, the country which has taken the experiment further than any other. The maximum penetration that is possible, due to the uncontrollable output of wind turbines, means that they could contribute at best 10% of U.S. energy demand. Even if per capita energy demand remains constant, that 10% would be cancelled out by U.S. population growth in 10 years. In summary, installing wind turbines will not keep up with the present U.S. population growth, let alone give a bulwark of energy security to the present population. However, the whole situation, for wind and other uncontrollables, will need reviewing if compressed air electrical storage, CAES, is shown — even in some countries and the USA is a promising one — to be a practical proposition.

7

Renewable Diesel

Concerns about the environmental impact of fossil fuels – as well as the possibility that fossil fuel production may soon fall short of demand – have spurred a search for renewable alternative fuels. Distillates, the class of fossil fuels which includes diesel and fuel oil, account for a significant fraction of worldwide fossil fuel demand. Renewable distillates may be produced via several different technologies and from a wide variety of raw materials. Renewable distillates may be categorized as biodiesel, which is a mono-alkyl ester and not a hydrocarbon, or 'green diesel', which is a renewable hydrocarbon diesel produced via either hydrotreating or biomass to liquids (BTL) technology. There are, however, important ecological and economic tradeoffs to consider. While the expansion of renewable diesel production may provide additional sources of income for farmers in tropical regions, it also provides economic incentive for clearing tropical forests and negatively impacting biodiversity. Also, many of the raw materials used to produce renewable diesel are edible, or compete with arable land used to grow food. This creates potential conflicts over the use of biomass for food or for fuel. In contrast to first-generation renewable diesel technologies which utilize primarily edible oils, BTL technology can utilize any type of biomass for diesel production. However, high capital costs have thus far hampered development of BTL technology.

INTRODUCTION

Distillate fuel oils, a category of fuels which includes petroleum diesel and home heating oil, account for almost 30% of worldwide petroleum consumption (EIA 2004). As fossil fuel reserves continue to deplete, sustainable alternatives to petroleum-based products are needed. One potential energy source is renewable distillate fuel oils produced from biomass. Such biofuels have a long history, as peanut oil and whale oil were used as lubricants and energy sources long before they were displaced by petroleum products.

Biomass-derived diesel substitutes can be produced via several different technologies and from a wide variety of starting materials. Renewable diesel may be produced from edible vegetable oils such as soybean oil, cottonseed oil, or rapeseed oil – non-edible oils such as jatropha oil or algal oils – animal fats, and even waste cooking grease. This chapter will examine the differences between various renew ble diesel technologies, the variety of raw materials that can be used to produce renewable diesel, as well as possible trade-offs involved in wide-scale adoption of these alternatives.

THE DIESEL ENGINE

The advantages of using distillates as a fuel source go beyond the fact that distillates and their substitutes are typically more energy dense than gasoline and gasoline substitutes. The diesel, or compression-ignition engine (CIE) is different from a gasoline engine, or spark-ignition engine (SIE) in several respects. Whereas the SIE is normally ignited by a spark plug, the CIE is ignited by compression. The CIE achieves a much higher compression ratio,1 which allows for a more powerful combustion, thus enabling more useful work to be realized. The result is that the efficiency of the CIE is up to 40% greater than for an SIE. Therefore, on purely the basis of engine efficiency, the CIE and fuels that can run in a CIE are preferred. A fuel must be resistant to ignition as it is being compressed if it is to be considered as an appropriate fuel for a CIE. Gasoline does not fall into this category, which is why it is not used in CIEs. But diesel fuels do fall into this category. Diesel substitutes produced from biomass are the subject of this chapter.

ECOLOGICAL LIMITS

Before examining potential renewable distillates, consider the question: What is the potential of biofuels with respect to ending the world's petroleum dependence? If biofuels are to make a meaningful dent in present worldwide oil usage of around 85 million barrels per day, then a massive expansion from current production capacity would be required. For example, as of this writing U.S. production of ethanol – seven billion gallons per year – is less than the energy equivalent of 1% of U.S. oil consumption.2 Yet this is purely on a gross basis, which presumes that there are no petroleum inputs into the production of ethanol. Because fossil fuels are used to grow and harvest corn, and then to operate the ethanol distillery, the net energy added to the U.S. energy supply is much smaller. Yet even this negligible contribution to energy supplies is arguably resulting in a number of undesirable consequences.

But even ignoring the potential negatives, can one presume that biofuels can make a significant contribution to present energy demands? Consider the following thought experiment. There are 148.94 million square kilometers of land area in the world, 13.31% of which are considered to be arable (CIA 2007). Permanent crops occupy 4.71% of the total land area, leaving 12.8 million square kilometers (1.28 billion hectares) of arable land potentially (for the purpose of the thought experiment) available for cultivation of biofuels.3 There are many different feed stocks from which to make renewable diesel, but most of the world's biodiesel is made from rapeseed oil (Puppan 2002). Rapeseed is an oilseed crop that is widespread and produces relatively high oil production. Unlike ethanol, which has an energy content 1/3rd less than that of gasoline, rapeseed oil has an energy density closer to that of petroleum.

Consider how much petroleum might be displaced if all 1.28 billion hectares of arable land were planted in rapeseed, or an energy crop with an oil productivity similar to rapeseed. While the average worldwide yield is substantially lower, rapeseed growers in Germany have succeeded in pushing oil yields to 2.9 tons/ha (Puppan 2002). If the rest of the world could achieve these high levels, this would result in a hypothetical worldwide oil yield of 3.7 billion tons. The energy content of rapeseed oil is about 10% less than that of petroleum diesel, so the gross petroleum equivalent yield from this exercise is 3.3 billion tons per year.

Because it takes energy to produce the biomass and process into fuel, the net yield will be lower, and in some cases may even be negative (i.e., more energy put into the process than is contained in the final product). Lewis compared several studies that examined the energy inputs required to produce biodiesel from rapeseed (Lewis 1997). Depending on the assumptions made, the energy input estimates ranged from 0.382 to 0.870 joules of input per joule of biodiesel produced and distributed. Assuming the best case value (lowest energy inputs) of 0.382, the net petroleum equivalent yield of rapeseed oil is reduced to 2 billion tons per year.4 The world's present usage of petroleum, 85 million barrels per day, is equivalent to 4.25 billion metric tons per year. By making very optimistic assumptions on the amount of land devoted to biofuels, the oil yield per hectare, and the energy inputs to produce the biofuels, the net is still less than half of the world's current demand for petroleum.

Of course this is merely a thought experiment. Positive and negative externalities (e.g., the potential impact on food prices on one hand; the income opportunities for 3rd world farmers on the other) have been ignored. There are many considerations that could influence the result in one direction or another. But the exercise highlights the difficulty the world would face in attempting to replace our petroleum usage with biofuels.

STRAIGHT VEGETABLE OIL

Unmodified vegetable-derived triglycerides, commonly known as vegetable oil, may be used to fuel a diesel engine. Rudolf Diesel demonstrated the use of peanut oil as fuel for one of his diesel engines at the Paris Exposition in 1900 (Altin et al. 2001). Modern diesel engines are also capable of running on straight (unmodified) vegetable oil (SVO) or waste grease, with some loss of power over petroleum diesel (West 2004). Numerous engine performance and emission tests have been conducted with SVO derived from many different sources, either as a standalone fuel or as a mixture with petroleum diesel (Fort and Blumberg 1982, Schlick et al. 1988, Hemmerlein et al. 1991, Goering et al. 1982).

The advantage of SVO as fuel is that a minimal amount of processing is required, which lowers the production costs of the fuel. The energy return for SVO, defined as energy output over the energy required to produce the fuel, will also be higher due to the avoidance of energy intensive downstream processing steps. There are several disadvantages of using SVO as fuel. The first is that researchers have found that engine performance suffers, and that hydrocarbon and carbon monoxide emissions increase relative to petroleum diesel. Particulate emissions were also observed to be higher with SVO. However, the same studies found that nitrogen oxide (NOx) emissions were lower for SVO (Altin et al. 2001). On longterm tests, carbon deposits have been found in the combustion chamber, and sticky gum deposits have occurred in the fuel lines (Fort and Blumberg 1982). SVO also has a very high viscosity relative to most diesel fuels. This reduces its ability to flow, especially in cold weather. This characteristic may be compensated for by heating up the SVO, or by blending it with larger volumes of lower viscosity diesel fuels.

BIODIESEL

Definition

Biodiesel is defined as the mono-alkyl ester product derived from lipid5 feedstock like SVO or animal fats (Knothe 2001). The chemical structure is distinctly different from petroleum diesel, and biodiesel has somewhat different physical and chemical properties from petroleum diesel.

Biodiesel is normally produced by reacting triglycerides (long-chain fatty acids contained in the lipids) with an alcohol in a base-catalyzed reaction (Sheehan 1998). Methanol, ethanol, or even longer chain alcohols may be used as the alcohol, although lower-cost and faster-reacting methanol6 is typically preferred. The primary products of the reaction are the alkyl ester (e.g., methyl ester if methanol is used) and glycerol. The key advantage over SVO is that the viscosity is greatly reduced, albeit at the cost of additional processing and a glycerol byproduct.

BIODIESEL CHARACTERISTICS

Biodiesel is reportedly nontoxic and biodegradable (Sheehan et al. 1998). An EPA study published in 2002 showed that the impact of biodiesel on exhaust emissions was mostly favorable

(EPA 2002). Compared to petroleum diesel, a pure blend of biodiesel was estimated to increase the emission of NOx by 10%, but reduce emissions of carbon monoxide and particulate matter by almost 50%. Hydrocarbon emissions from biodiesel were reduced by almost 70% relative to petroleum diesel. However, other researchers have reached different conclusions. While confirming the NOx reduction observed in the EPA studies, Altin et al. determined that both biodiesel and SVO increase CO emissions over petroleum diesel (Altin et al. 2001). They also determined that the energy content of biodiesel and SVO was about 10% lower than for petroleum diesel. This means that a larger volume of biodiesel consumption is required per distance traveled, increasing the total emissions over what a comparison of the exhaust concentrations would imply.

The natural cetane7 number for biodiesel in the 2002 EPA study was found to be higher than for petroleum diesel (55 vs. 44). Altin et al. again reported a different result, finding that in most cases the natural cetane numbers were lower for biodiesel than for petroleum diesel. These discrepancies in cetane results have been attributed to the differences in the quality of the oil feedstock, and to whether the biodiesel had been distilled (Van Gerpen 1996).

A major attraction of biodiesel is that it is easy to produce. An individual with a minimal amount of equipment or expertise can learn to produce biodiesel. With the exception of SVO, production of renewable diesel by hobbyists is limited to biodiesel because a much larger capital expenditure is required for other renewable diesel technologies. Biodiesel does have characteristics that make it problematic in cold weather conditions. The cloud and pour points8 of biodiesel can be 20°C or higher than for petroleum diesel (Kinast 2003). This is a severe disadvantage for the usage of biodiesel in cold climates, and limits the blending percentage with petroleum diesel in cold weather.

ENERGY RETURN

The energy return of biodiesel is disputed. Sheehan et al. reported in 1998 that the production of 1 megajoule (MJ) of soy-derived biodiesel required 0.3110 MJ of fossil fuel inputs, for a fossil energy ratio9 of 3.2 (Sheehan et al. 1998). They further reported that during the production of biodiesel from soybeans, the soybean crushing and soybean conversion steps required the most energy, respectively using 34.25% and 34.55% of the total energy. The remainder of the energy inputs came mostly from agriculture, at approximately 25% of the total energy input. However, Pimentel and Patzek reported that the energy return for soy biodiesel is slightly less than 1.0, meaning that soy biodiesel is nonrenewable according to their study (Pimentel and Patzek 2005). But there were some differences in the methodology employed. The two studies allocated energy differently between the soy oil product and the soy meal product. This resulted in very different energy input calculations. Sheehan assigned to the soy oil a fossil energy input from the agricultural step equivalent to 0.0656 MJ per MJ of biodiesel produced. Pimentel and Patzek assigned an energy input from the agricultural step equivalent to 0.70 MJ per MJ of biodiesel produced – over 10 times the amount from the Sheehan study.10 However, the Pimentel and Patzek study found that the energy return from the soybean cultivation step was renewable (considering only energy inputs), with 2.56 MJ of soybeans being returned for an energy input of 1.0 MJ.

Glycerol Byproduct

One of the challenges in the production of biodiesel is disposal of the glycerol11 byproduct. This has created such a glut of glycerol, that some glycerol producers have been forced to shut down plants (Boyd 2007). Excess glycerol is currently disposed of by incineration, prompting the UK's Department for Trade and Industry to fund projects exploring the conversion of glycerol into value-added chemicals (Glycerol Challenge 2007).

GREEN DIESEL

Definition

Another form of renewable diesel is 'green diesel.' Green diesel is chemically the same as petroleum diesel, but it is made from recently living biomass. Unlike biodiesel, which is an ester and has different chemical properties from petroleum diesel, green diesel is composed of long-chain hydrocarbons, and can be mixed with petroleum diesel in any proportion for use as transportation fuel. Green diesel technology is frequently referred to as second-generation renewable diesel technology. There are two methods of making green diesel. One is to hydroprocess vegetable oil or animal fats. Hydroprocessing may occur in the same facilities used to process petroleum. The second method of making green diesel involves partially combusting a biomass source to produce carbon monoxide and hydrogen – syngas – and then utilizing the Fischer-Tropsch reaction to produce complex hydrocarbons. This process is commonly called the biomass-to-liquids, or BTL process.

Hydroprocessing

Hydroprocessing is the process of reacting a feed stock with hydrogen under elevated temperature and pressure in order to change the chemical properties of the feed stock. The technology has long been used in the petroleum industry to 'crack', or convert very large organic molecules into smaller organic molecules, ranging from those suitable for liquid petroleum gas (LPG) applications through those suitable for use as distillate fuels.

In recent years, hydroprocessing technology has been used to convert lipid feed stocks into distillate fuels. The resulting products are a distillate fuel with properties very similar to petroleum diesel, and propane (Hodge 2006). The primary advantages over first-generation biodiesel technology are:

(1) The cold weather properties are superior;

(2) The propane byproduct is preferable over glycerol byproduct;

(3) The heating content is greater;

(4) The cetane number is greater; and

(5) Capital costs and operating costs are lower (Arena et al. 2006).

A number of companies have announced renewable diesel projects based on hydroprocessing technology. In May 2007 Neste Oil Corporation in Finland inaugurated a plant that will produce 170,000 t/a of renewable diesel fuel from a mix of vegetable oil and animal fat (Neste 2007). Italy's Eni has announced plans for a facility in Livorno, Italy that will hydrotreat vegetable oil for supplying European markets. Brazil's Petrobras is currently producing renewable diesel via their patented hydrocracking technology (NREL 2006). And in April 2007 ConocoPhillips, after testing their hydrocracking technology to make renewable diesel from rapeseed oil in Whitegate, Ireland, announced a partnership with Tyson Foods to convert waste animal fat into diesel (ConocoPhillips 2007).

Like biodiesel production, which normally utilizes fossil fuel-derived methanol, hydroprocessing requires fossil fuel-derived hydrogen.12 No definitive life cycle analyses have been performed for diesel produced via hydroprocessing. Therefore, the energy return and overall environmental impact have yet to be quantified.

Biomass-to-Liquids

When an organic material is burned (e.g., natural gas, coal, biomass), it can be completely oxidized (gasified) to carbon dioxide and water, or it can be partially oxidized to carbon monoxide and hydrogen. The latter partial oxidation (POX), or gasification reaction, is accomplished by restricting the amount of oxygen during the combustion. The resulting mixture of carbon monoxide and hydrogen is called synthesis gas (syngas) and can be used as the starting material for a wide variety of organic compounds, including transportation fuels.

Syngas may be used to produce long-chain hydrocarbons via the Fischer-Tropsch (FT) reaction. The FT reaction, invented by German chemists Franz Fischer and Hans Tropsch in the 1920s, was used by Germany during World War II to produce synthetic fuels for their war effort. The FT reaction has received a great deal of interest lately because of the potential for converting natural gas, coal, or biomass into liquid transportation fuels. These processes are respectively referred to as gas-to-liquids (GTL), coal-to-liquids (CTL), and biomass-to-liquids (BTL), and the resulting fuels are 'synthetic fuels' or 'XTL fuels'. Of the XTL processes, BTL produces the only renewable fuel, as it utilizes recently anthropogenic (atmospheric) carbon.

Renewable diesel produced via BTL technology has one substantial advantage over biodiesel and hydrocracking technologies: Any source of biomass may be converted via BTL.

Biodiesel and hydrocracking processes are limited to lipids. This restricts their application to a feedstock that is very small in the context of the world's available biomass. BTL is the only renewable diesel technology with the potential for converting a wide range of waste biomass. Like GTL and CTL, development of BTL is presently hampered by high capital costs. According to the Energy Information Administration's Annual Energy Outlook 2006, capital costs per daily barrel of production are $15,000–20,000 for a petroleum refinery, $20,000–$30,000 for an ethanol plant, $30,000 for GTL, $60,000 for CTL, and $120,000–$140,000 for BTL (EIA 2006).

While a great deal of research, development, and commercial experience has gone into FT technology in recent years,13 biomass gasification technology is a relatively young field, which may partially explain the high capital costs. Nevertheless, the technology is progressing. Germany's Choren is building a plant in Freiberg, Germany to produce 15,000 tons/yr of their SunDiesel_ product starting in 2008 (Ledford 2006).

Feed Stocks

While renewable diesel may be produced from a wide variety of feed stocks, this section will focus on those that are either in widespread use, or are frequently discussed as feed stocks with very high potential for producing biofuels. Feed stocks for the BTL process will not be discussed, as any biomass source can be used for this process. The following feed stocks are specific to the lipid conversion technologies discussed in this chapter.

Soybeans

The United States is the world's largest producer of soybean oil (Sheehan 1998), producing approximately 10 million metric tons in 2006 (USDA June 2007).Worldwide production of soybean oil is 35 million metric tons (Rupilius and Ahmad 2007). Soybean oil is typically produced by cracking the soybeans and extracting the oil with a solvent such as hexane. Finished soybean oil is widely used as cooking oil, in various processed foods, and for the production of biodiesel. Relative to other oil crops, productivity of oil from soybeans is low. Soybean yields in 2006 in the U.S. amounted to 2871 kg/ha (USDA January 2007). At a typical soybean oil yield of 18%, this would have produced an average oil yield of 0.52 tons/ha. The average yield in Brazil, another major producer of soybean oil, has been reported at 0.40 tons/ha. These oil yields are far below reported yields of other oil crops such as rapeseed, palm oil, or coconut.

While the oil yields are low, soybean oil does have an advantage over many bio-oil crops. Soybeans are capable of atmospheric nitrogen fixation, so they can be grown with little or no nitrogen fertilizer inputs (Pimentel and Patzek 2005). Because nitrogen-based fertilizers are energy intensive to produce, the energy balance for the agricultural step should be much more favorable than for crops requiring nitrogen fertilizer. This also means that soybeans will contribute less water pollution in the way of fertilizer runoff into waterways.

The expansion of soybean cultivation is not without controversy. In Brazil, critics have charged that soybean cultivation is a major driver of deforestation in Amazonia, resulting in multiple negative impacts on biodiversity (Fearnside 2001). Some researchers also argue that the potential for drought is increasing due to the increased reflectivity of the cleared land (Costa et al. 2007). In the United States, use of genetically-modified soybeans is common. This has resulted in criticism from various countries and environmental groups opposed to the practice.

Rapeseed

Whereas biodiesel in the U.S. is produced primarily from soybean oil, rapeseed oil, also sometimes called canola,15 is the feedstock of choice for European biodiesel (Thuijl et al. 2003). Like soybean oil, rapeseed oil is edible. Rapeseed oil yields are about 1 ton/ha – double those of soybean oil. Rapeseed is produced mainly in China, Canada, the Indian subcontinent, and Northern Europe (Downey 1990). Rapeseed oil was the first vegetable oil used for transesterification to biodiesel, and remains the most widely-utilized vegetable oil in the production of biodiesel (Puppan 2002). The most common biodiesel produced from rapeseed oil is called Rapeseed-Methyl- Ester, or RME. RME has a slightly higher energy density than most biodiesels, and produces lower NOx and CO emissions than biodiesel produced from soybean oil

(EPA 2002). The primary disadvantage of rapeseed relative to some oil crops is that it has high nitrogen fertilizer requirements. Some life cycle analyses have shown a relatively small environmental benefit from RME relative to petroleum diesel, and a higher energy input than soybean oil, primarily because of the fertilizer requirements (De Nocker et al. 1998, Zemanek and Reinhardt 1999).

Palm Oil

Palm oil is an edible oil extracted from the fruit of the African Oil Palm. In 2006, worldwide palm oil production surpassed soybean oil to become the most widely produced vegetable oil in the world. In 2006, palm oil production was 37 million tons and accounted for just over 25% of all biological oil production (Rupilius and Ahmad 2007). This is a substantial oil yield relative to other lipid crops. For perspective, total distillate usage (diesel and fuel oil) in the United States was approximately 208.5 million tons16 in 2006 (EIA 2007).

By far the most productive lipid crop, palm oil is the preferred oil crop in tropical regions. The yields of up to five tons of palm oil per hectare can be ten times the per hectare yield of soybean oil (Mattson et al. 2000). Palm oil is a major source of revenue in countries like Malaysia, where earnings from palm oil exports exceed earnings from petroleum products (Kalam and Masjuki 2002). Palm oil presents an excellent case illustrating both the promise and the peril of biofuels. Driven by demand from the U.S. and the European Union (EU) due to mandated biofuel requirements, palm oil has provided a valuable cash crop for farmers in tropical regions like Malaysia, Indonesia, and Thailand. The high productivity of palm oil has led to a dramatic expansion in most tropical countries around the equator (Rupilius and Ahmad 2007). This has the potential for alleviating poverty in these regions.

But in certain locations, expansion of palm oil cultivation has resulted in serious environmental damage as rain forest has been cleared to make room for new palm oil plantations. Deforestation in some countries has been severe, which negatively impacts sustainability criteria, because these tropical forests absorb carbon dioxide and help mitigate global warming (Schmidt 2007).

Destruction of peat land in Indonesia for palm oil plantations has reportedly caused the country to become the world's third highest emitter of greenhouse gases (Silvius et al. 2006).

As a result of the potential environmental dangers posed by the expansion of biofuels, the Dutch government is developing sustainability criteria for biomass that will be incorporated into relevant policy decisions (Cramer 2006). The intention is employ life cycle analyses (LCAs) to measure the overall impact from using various biomass sources. For instance, if the developed world mandates large amounts of biofuels, but this come at the price of massive deforestation of tropical rainforests, the LCA will attempt to incorporate those negatives into the overall assessment. The categories that the Dutch group intends to evaluate are:

(1) Greenhouse gas balance;
(2) Competition with food, local energy supply, medicines and building materials;
(3) Biodiversity;
(4) Economic prosperity;
(5) Social well-being; and
(6) Environment.

In addition to the Dutch initiative, some other countries are evaluating the sustainability of biofuels (Rollefson et al. 2004). Yet such efforts may be ultimately futile unless a binding, worldwide agreement can be implemented. While slash-and-burn growers may find that the Dutch will not buy their products, they may easily find other buyers for their product in the global marketplace.

Jatropha

Jatropha curcas is a non-edible shrub native to tropical America, but now found throughout tropical and subtropical regions of Africa and Asia (Augustus et al. 2002). Jatropha is well-suited for growing in arid conditions, has low moisture requirements (Sirisomboon et al. 2007), and may be used to reclaim marginal, desert, or degraded land (Wood 2005). The oil content of the seeds ranges from 30% to 50%, and the unmodified oil has been shown to perform adequately as a 50/50 blend with petroleum diesel (Pramanik 2003). However, as is the case with other bio-oils, the viscosity of the unmodified oil is much higher than for petroleum diesel. The heating value and cetane number for jatropha oil are also lower than for petroleum diesel. This means it is preferable to process the raw oil into biodiesel or green diesel. Jatropha appears to have several advantages as a renewable diesel feedstock. Because it is both non-edible and can be grown on marginal lands, it is potentially a sustainable biofuel that will not compete with food crops. This is not the case with biofuels derived from soybeans, rapeseed, or palm.

Jatropha seed yields can vary over a very large range – from 0.5 tons per hectare under arid conditions to 12 tons per hectare under optimum conditions (Francis et al. 2005). However, if marginal land is to be used, then yields in the lower range will probably by typical. Makkar et al. determined that the kernel represents 61.3% of the seed weight, and that the lipid concentration represented 53.0% of the kernel weight (Makkar et al. 1997). Therefore, one might conservatively estimate that the average oil yield per hectare of jatropha on marginal, non-irrigated land may be 0.5 tons times 61.3% times 53.0%, or 0.162 tons of oil per hectare. Jatropha oil contains about 90% of the energy density of petroleum diesel, so the energy equivalent yield is reduced by an additional 10% to 0.146 tons per hectare. While this is substantially less than the oil production of soybeans, rapeseed, or palm oil, the potential for production on marginal land may give jatropha a distinct advantage over the higherproducing oil crops.

A commercial venture was announced in June 2007 between BP and D1 Oils to develop jatropha biodiesel (BP 2007). The companies announced that they will invest $160 million with the stated intent of becoming the largest jatropha biodiesel producer in the world. The venture intends to produce volumes of up to 2 million tons of biodiesel per year. Jatropha has one significant downside. Jatropha seeds and leaves are toxic to humans and livestock. This led the Australian government to ban the plant in 2006. It was declared an invasive species, and 'too risky forWestern Australian agriculture and the environment here' (DAFWA 2006).

While jatropha has intriguing potential, a number of research challenges remain. Because of the toxicity issues, the potential for detoxification should be studied (Heller 1996). Furthermore, a systematic study of the factors influencing oil yields should be undertaken, because higher yields are probably needed before jatropha can contribute significantly to world distillate supplies.17 Finally, it may be worthwhile to study the potential for jatropha varieties that thrive in more temperate climates, as jatropha is presently limited to tropical climates.

ALGAE

Certain species of algae are capable of producing lipids, which can be pressed out and then converted to renewable diesel. Algae-based renewable diesel is an appealing prospect, as this could potentially open up biofuel production to areas unsuitable for farming. Furthermore, the estimates of the oil production potential from algae have been as high as 160 tons/ha – 30 times that of palm oil. From 1978 to 1996, the U.S. Department of Energy funded a study by the National Renewable Energy Laboratory (NREL) on the feasibility of producing renewable fuels from algae (Sheehan et al. 1998). The study examined a number of strains of algae for potential lipid production, as well as those that could grow under conditions of extreme temperature, pH, and salinity. Researchers examined the molecular biology and genetics of algae, and identified important metabolic pathways for the production of lipids.

While the production of biofuels from a raw material like algae has obvious appeal, the NREL close-out report concluded that there are many technical challenges to be overcome. A major challenge was encountered in the attempts to increase oil yields. Oil concentrations could be increased by stressing the algae and causing it to shift from a growth mode into a lipid production mode, but this resulted in lower overall oil yields because algal growth slowed. The researchers also discovered that contamination was often a problem upon moving from the laboratory into open pond systems.

The close-out report suggested that algae could potentially supply the equivalent of a large fraction of U.S. demand, but costs must come down, and technical challenges must be solved. On the subject of costs, the report noted 'Even with aggressive assumptions about biological productivity, we project costs for biodiesel which are two times higher than current petroleum diesel fuel costs.' Furthermore, because of lack of data on continuous lipid production from algae, the energy return on the process is unknown.

Animal Fats

Total production of animal fats in the U.S. was approximately 4.5 million tons in 2006 (U.S. Census Bureau 2007). This is just under half the mass of soybean oil produced each year in the U.S. It is also the energy equivalent of around 1.5 days of U.S. petroleum demand.

Animal fats contain fewer double bonds than do most vegetable oils (Peterson 1986). This has an influence on the properties of the renewable diesel product. For example, biodiesel properties have been shown to vary depending on whether the biodiesel was produced from animal or plant lipids. In 2002, the EPA compared plant-based biodiesels derived from soybean, rapeseed, and canola oils, to animal-based biodiesels derived from tallow, grease, and lard (EPA 2002).

The study found that animal-based biodiesels had a slightly lower energy density, but higher cetane numbers than plant-based biodiesels. The study also found that animal-based biodiesel produced substantially fewer NOx and particulate matter emissions.

Animal fats also respond differently to the hydrotreating process than do vegetable oils. Animal fats are more amenable to the hydrotreating process because double bonds are saturated in the hydrotreating process. Feed stocks like animal fats, with fewer double bonds than vegetable oils, will require less hydrogen to convert the oil to green diesel.

While animal fats are a byproduct of meat processing, there are significant environmental costs associated with industrial animal agriculture. The production of meat is a highly inefficient process. The production of beef requires relatively large inputs of water, grain, forage, and fossil fuels. Production of 1 kilocalorie of beef protein requires a fossil fuel input of 40 kilocalories (Pimentel and Pimentel 2003). This suggests that animal-based biofuels may be legitimately considered recycled fossil fuels.

Waste Biomass

North America and Western Europe combine to produce an estimated 500 million tons of municipal waste (UNEP 2004a).

The main contributors to municipal waste throughout the developed world are organic materials such as food waste, grass clippings, waste cooking oils, and paper (UNEP 2004b).Waste biomass that is presently destined for landfills has great appeal as a feedstock for biofuels production, as it is an available biomass source that does not compete with food.

Of this waste biomass, the BTL process can potentially convert any of it to liquid fuels. The lipid conversion technologies are however limited to the waste cooking oil fraction.

Waste cooking oils can either be converted to biodiesel via transesterification, or to green diesel via hydrotreating. For the hobbyist, the waste oil feedstock can often be acquired from restaurants at little or no cost.

The conversion to biodiesel may be carried out without expending a great deal of capital, meaning that biodiesel can be produced from waste cooking oil at a very low cost.

Businesses are beginning to realize the opportunity in recycling waste cooking oil into transportation fuel.

In July 2007, McDonald's UK restaurants announced their intention to run their delivery fleet on the waste cooking oil generated by 900 of their restaurants (McDonald's 2007). A program under way in New York City is on pace to recycle 450 tons of used cooking oil to biodiesel in 2007 (RWA 2007).

CONCLUSIONS

Biofuels can contribute to our energy portfolio, and many different options are available. But some options pose high environmental risks, some compete with food, and some are far more sustainable than others.

Each option should be carefully weighed against the overall impact on the environment and society as a whole. Sustainable energy solutions must be pursued, and rigorous life cycle analyses should be undertaken for all of our energy choices.

We live in a world with limited resources, and a declining endowment of fossil fuel reserves. Much of the world aspires to a higher standard of living. The energy policies that we pursue should attempt to balance the needs of all citizens, worldwide. These policies must carefully consider the ecology of the planet, so future generations are not denied opportunities because of the choices we make today.

Conversion Factors and Calculations

While SI units are used in this chapter, Imperial/UK units are commonly used in the UK and in the U.S. Therefore, a number of common conversion factors are listed here which should enable to reader to convert between SI and Imperial units. A number of measures in the text have been converted from Imperial units, but the conversion factors listed should enable the reader to reproduce all figures.

Also, because different assumptions of physical properties (density, energy content, etc.) will lead to slightly different results, certain assumptions and calculations used in this chapter are provided in this section.

Conversion Factors

1 barrel of oil = 42 gallons = 158.984 liters = 0.137 metric tons

1 barrel of oil = 5.8 million BTUs of energy = 6.1 gigajoules (GJ)

1.0 hectare = 10,000m2 = 2.47 acres

The specific gravity of crude oil is 0.88.

The specific gravity of diesel oils is 0.84.

The specific gravity of biodiesel is 0.88.

The specific gravity of ethanol is 0.79.

Lower Heating Values

The lower heating value (LHV) is the heat released by combusting a substance without recovering the heat lost from vaporized water. The LHV is a more accurate representation of actual heat utilized during combustion, as vaporized water is rarely recovered.

The LHV for crude oil is 138,100 Btu/gallon = 38.5MJ/liter ≐ 45.3GJ/t

The LHV for distillates is 130,500 Btu/gallon = 36.4MJ/liter = 42.8GJ/t

The LHV for biodiesel is 117,000 Btu/gallon = 32.6MJ/liter = 37.8GJ/t

The LHV for ethanol is 75,700 Btu/gallon = 21.1MJ/liter = 26.7GJ/t

Calculations: In this section, several of the calculations referenced in the text are reproduced. Calculation 1: Current oil usage in the United States is approximately 21 million barrels per day. The energy value of 1 barrel of oil is approximately 5.8 million BTUs.

Ethanol production of 7 billion barrels per year is equivalent to 457,000 barrels per day. This is 2.2% of daily oil usage on a volumetric basis, but ethanol has approximately 76,000 BTUs/bbl, versus 138,000 BTUs/bbl for oil.

Therefore, 7 billion gallons of ethanol per year is worth 1.2% of U.S. daily oil consumption. Backing out the energy inputs required to produce the ethanol (fossil fuels for tractors, trucking, fertilizer, pesticides, etc.) drops the net offset to well less than 1% of U.S. daily oil consumption.

Calculation 2: If the energy input is 0.382, then the net energy is (1-0.382) " 3.3 billion tons of rapeseed oil.

The balance of 1.26 billion tons would be equivalent to the energy required to produce, process, and distribute the final product. Calculation 3: In the United States, distillate demand in 2006 was 4.17 million barrels per day. One barrel of oil is equivalent to 0.137 metric tons; therefore distillate demand in 2006 was 0.57 tons per day, or 208.5 tons per year.

Calculation 4: Consider the potential for displacing 10% of the world's distillate demand of 1.1 billion tons per year – 110 million tons - with jatropha oil.

Jatropha, with about 10% less energy than petroleum distillates, will require 122 million tons (110 million/0.9) on a gross replacement basis (i.e., not considering energy inputs). On marginal, un-irrigated land the yields will likely be at the bottom of the range of observed yields. At a yield of 0.146 tons per hectare, this would require 836 million hectares, which is greater than the 700 million hectares currently occupied by permanent crops. An estimated 2 billion acres is considered to be degraded and perhaps suitable for jatropha cultivation (Oldeman et al. 1991).

There are also an estimated 1.66 billion hectares in Africa that are deemed suitable for jatropha production (Parsons 2005). This could provide a valuable cash crop for African farmers. But, until an estimate is made of the energy inputs required to process and distribute the jatropha-derived fuel on a widespread basis – especially on marginal land – the real potential for adding to the world's net distillate supply is unknown.

8

Biomass Fuel Cycle Boundaries and Parameters: Current Practice and Proposed Methodology

A methodology is presented for standardizing Biomass Fuel Cycle (BFC) analysis and evaluation. The Biomass Fuel Cycle Methodology (BFCM) enables eliminating disparities, minimizing differences, and clearly quantifying variations. Standardized templates, modular staging, and normalized analysis formulations are used to disposition technologies, facilities, activities, boundaries, and parameters. The methodology enables presentation of quantification and characterization information in a straightforward standard format applicable across a broad range of BFC's. BFC literature data is used to illustrate the flexibility, clarity, and diversity of the methodology. The types of insights to be gained concerning the limitations of BFC treatments (boundary shortcomings, energy uncertainties, analysis constraints) are discussed.

INTRODUCTION

The US national security driven: energy independence goal, reduction of pollution, and the pursuit of renewable energy source efforts have resulted in government biofuels subsidies of $6 billion per year (Koplow, 2006), and industry development of Biomass Fuel Cycles (BFC's). A methodology has been developed to provide unbiased characterization and analysis for use in technology viability evaluation. The selection of the boundaries, parameters, and associated numerical values for a given BFC has a direct impact on the evaluation of that technology's viability, import to energy independence, and renewable energy value. Currently there are significant judgment differences about BFC component import, analysis scope, boundary selection, and parameter values. Opinions differ and modeled scopes vary on topics such as coproduct energy credit, facility fabrication, waste management, environmental, and parameter numerical value.

The biomass fuel cycle methodology (BFCM) presented is intended to assist in avoiding, minimizing, or, at least, clearly quantifying and delineating analysis differences. The BFCM uses templates, modular modeling, scenario definition, and statistical based methods to standardize analyses, establish unbiased boundary assignments, normalize numerical value treatments, treat data uncertainty, and characterize limitations of results. Adding clarity to the understanding of BFC intricacies and analyses is intended to facilitate national level discussions and decisions on development of biomass fuel capabilities such as infrastructure requirements for an expanded ethanol industry (Brent and Yacobucci, 2006). In the present study, the focus is on the energy and environmental aspects of BFC's.

BFC ANALYSIS METHODOLOGY: A MODULAR MODEL APPROACH

The BCFM is structured so as to be applicable to a broad range of BFC's. The methodology's three stage template system, fuel cycle parameters, boundary treatment, and statistical tools are presented. The approach facilitates modeling and analysis of scenarios involving diverse configurations (e.g., stand alone biomass cycles, crop rotation combined BFC's), agricultural variations (e.g., fertilization versus crop rotation, extent of tilling, silage practices/use), biomass to fuel processing variations (e.g., dry versus wet corn milling, cogeneration, cellulous digestion), energy balance consideration, and environmental impact assessment.

BFC General Stages and Templates

The BFCM structures each BFC analysis based on three main analysis stages: 1. Infrastructure – multi-user services/facilities:

70 Sub-activities (59 distinctive + 11 onsite waste management covering 4 waste steam types)

The three general templates detail BFC processes and practices using a Phase, Subphase, Activity, and Sub-activity component structure. These template baselines identify components without consideration of specific BFC potential significance. Component significance will vary both within and across BFC's. Using the templates, specific BFC modules are established and the cycle boundaries are delineated. Each BFC module Sub-activity is dispositioned (i.e., assigned a parameter/value or justified as not a consideration). Thus each module documents the specifics for use in quantifying and characterizing its' BFC. Introduction into a module of new BFC process/practice components or sub-activities to show desired detail is straightforward. This template module approach readily accommodates customization of components while ensuring a standard set of sub-activities is addressed. The module components are analyzed using the standardized analysis and documentation methodologies thereby enabling inter-BFC and intra-BFC comparison. Although not explicitly addressed, the BFCM could be applied to monetary, production, distribution, regulatory, national security, incentives, and subsidies evaluations through selective expansion of the level of detail in the general templates. Having BFC evaluations linked via these common general templates is advantageous from a continuity, comparison, and clarity perspective.

BFC Parameters and Associated Variability

The BFC variability arises from natural and technological causes. Weather (e.g., wet/dry, temperature, storm damage), location(e.g., farm: soil type/condition, crop disease/pests; biorefinery: infrastructure, economics), transport distance (e.g., from farm to storage/process facility, biofuel distribution distance), seed type, agricultural practice (e.g., crop rotation, fertilization, irrigation), fuel source mix used within cycle (e.g., coal, gas, oil, biomass), biomass type (e.g., corn, soybean, switchgrass), and biofuel process technology (e.g., corn dry/wet mill, cellulose breakdown process) are typical sources of variability. Such viabilities are addressed and quantified by using two different types of parameters. The first is the biomass yield parameters used to quantitatively track the following sources of variability:

Weather, location, seed type, agricultural practice: Crop Yield = Ycrop

Biomass type, biofuel manufacture process: Biofuel Process Yield = Ybfp

The second parameter type is the individual parameters unique to a given module Sub-activity. In the BFCM treatment, Ycrop and Ybfp variability relationships are examined separately from the pk values.

Biomass Yield Parameters

For a given BFC:

Ncrop to bfp = Ycrop A Ybfp

Here Ncrop to bfp is the BFC net fuel production, Ycrop is the agriculture stage biomass crop yield, A is the planted land area, and Ybfp is the biofuel production stage yield. Another BFC general yield and biofuel energy relationship is:

Ebiofuel = Ncorn to bfp UEfuel e

Here Ebiofuel is the BFC created biofuel energy and UEfuel e is the biofuel useable energy. Combining and rearranging these two equations:

Ebiofuel/A = Ycrop Ybfp UEbiofuel (10.1)

Ebiofuel/A is a measure of the BFC crop and biomass fuel production efficiency in creating the biofuel. This equation enables biofuel yield evaluation at both the local/regional and national fuel cycle production levels.

Clearly gains in crop and process yields mean higher biofuel energy per acre planted.

Template Parameters

For each template Activity, there is an assigned k value. This k value is used to index the pk value assigned to that Activity and it's associated Sub-activities. The pk value and it's uncertainty _k are specific numerical values used in the analysis. Consider, for example, in Template 1 under the Facilities Phase there is the Seed

Plant Sub-phase. It's assigned Activity and associated Sub-activities index value is k = 5. Therefore it's numerical values used in an analysis are assigned to the p5 and _5 parameter in the BFCM equations discussed here. The pk's are used to calculate the Smodule j value of interest: Smodule j = fj(pk) and the _k's are used to quantify the uncertainty (_j) associated with that Smodule j. The fj(pk) equations are typically simple summations for the

BFC's but can be any mathematical relationship. The detail for a given Smodule j is determined by the BFC scenario and associated module. Both the Smodule j value and its' _j are used to quantifying and characterizing the BFC.

The general relationship applicable to each module is:

SBFC = m_ j=1 Smodule j Uj Fj (10.2)

Here SBFC is the total value (e.g., energy, mass, volume) for the given BFC modeled scenario made up of m modules; Uj is the land area planted, Biorefinery processed biomass, or biofuel volume; and Fj is the scenario specified decimal fraction factor used to evaluate a Uj variation (Fj = 1 if Uj held constant).

BFC yields, pk's, and _k's values, which are annual numbers, are reported in various units in the literature. In order to sum the Smodule j 's, the data must be normalize to a common unit. In the current treatment the numerical values are normalized to Btu/Acre. The conversion factors used were: 948.452 Btu/MJ, 0.2520 Kcal/Btu, 3.7854 L/Gal, and 2.471 Acre/Ha. The Biorefinery pk values were normalized to Btu/Acre using each specific study crop and biofuel yields. The resultant Smodule 3 values are thus a function of these specific yields which introduces two sources of variability into the analysis.

BFC Boundaries

A fundamental consideration is the establishment of the given BFC boundaries.

The boundaries for a given BFC are established by using Templates 1, 2, and 3 as the starting point. The three templates cover a broader range of BFC aspects than typically addressed. Their level of Sub-activity breakout focuses on aspects needing explicate dispositioning. The Sub-activities encompass materials, components, and facilities starting from natural resources through fabrication and usage to disposal. The pk's quantify aspects such as raw material extraction (e.g., mining of coal and minerals, petroleum drilling), materials fabrication (e.g., steel, fuel, fertilizer, farm equipment), construction (e.g., facilities, roads), operation (e.g., farming, storage, processing, transporting), and waste management (e.g., discharges, emissions, equipment and facility replaced or decommissioned).

The dispositioning (i.e., inclusion or exclusion) of a pk is a boundary decision. The BFC modules enable capturing the justification, including quantification of the impact, of Sub-activity exclusion. However, Sub-activity exclusion can result in important differences between models. Inclusion has the advantages of simplifying the description, facilitating cross model comparison and evaluation, and minimizing the potential for underestimating (which is inherent to BFC's as a result of their cumulative parameter property).

The end result is the BFC Stage Sub-activities and boundary demarcations are clearly delineated and justified. And the pk and _k values are presented in a standard format.

BFC Fuel and Net Energy Balance Definitions

The BFC energy measure of interest is the Net Energy Balance (NEB):

NEB = Total BFC Energy Gain (EG) – Total BFC Energy Loss (EL) = TEG " TEL Concise definition of EG and EL facilitates BFCM boundary dispositioning, energy accounting, and consistency.

Fuel Energy Definitions

When calculating the NEB, the energy gain (i.e., creation of fuel or productive use of BFC biomass or biofuel) and loss (i.e., consumption/expending of non-BFC fuel or energy) accounting needs to be well defined. The energy independence and environmental national goals lead to replacement of fossil fuels (both foreign and domestic) with domestic biomass fuels. BFC energy accounting needs to address all energy consumptions. The BFC energy definitions that follow directly from the above considerations are:

EL = Energy Loss for given BFC = directly (e.g., burned at given BFC facility) or indirectly (e.g., resource extraction/ production/refinement, electricity generation, steam generation, transport) expended fossil (i.e., petroleum, coal) fuels, biomass/ biofuel, electricity, or energy (e.g., heat) via nuclear/solar/ water/ wind power.

EG = Energy Gain for given BFC = created biofuels productive combustion (e.g., ethanol fuel oxidant in gasoline, ethanol replacement of gasoline, biodiesel replacement of petroleum diesel) + biomass or BFC created coproducts combustion supplying productive heat and/or power (e.g., silage, bagasse) + biomass, biofuels, or coproduct conversion to products (e.g., biomass digestion resulting in fertilizers, silage composting resulting in lowered field fertilization, conversion of biofuel to pesticides) that displace corresponding products derived from fossil (i.e., petroleum, coal) fuel.

Note both EL and EG include biomass/biofuel used to supply energy to the given BFC. The inclusion in both is needed in order to have the actual total energy value tabulated for the TEL and TEG. In this way both the TEL and TEG values are comprehensive and unencumbered with BFC specific exceptions/treatments. The accounting of the gain resulting from consumed biomass/biofuel displacing fossil fuel is captured in the EG analysis.

These definitions provide the basis for: excluding through definition the solar energy absorbed in growing the biomass and the caloric energy expended by BFC labor; retention of coproduct energy within the cycle unless some portion of the energy expended to create the coproduct is productively recovered by combustion of the coproduct; treating the use of solid, liquid, or gaseous biomass or biofuel within a given BFC as equivalent to an energy gain (i.e., those biomass fuel consumptions avoid consuming fossil fuels); and treating cogeneration as equivalent to an energy gain (i.e., it avoids consuming fossil fuels).

Fuel Useable Energy

The combustion of a fuel can be simplistically viewed as resulting in energy generation, water (as a gas) containing energy in the form of steam heat, combustion products, and particulates.

For fossil, biomass, and biofuel fuels, the relevant energy value is the usable energy realized when a quantity of fuel is burned under normal use conditions:

UE = Useable Energy = fuel High Heat Value (HHV) adjusted for normal use losses (L). HHV is also referred to as the gross heat content of a fuel. Combustion systems differ in their L value due to inefficiencies (e.g., heat leaks, energy transfer, discharge, friction) and operational variations.

For internal combustion engines it is typically assumed the efficiency is the same for all liquid fuels and the main loss is via steam. This L adjusted HHV is commonly referred to as the Low Heat Value (LHV) for the fuel (also called the net heat content) and is commonly used as the UE value. Use of the LHV provides a consistent, common base of comparison. Productive use of L, such as preheater use of boiler system exhaust, increases the UE value with respect to the LHV.

For combustion of solid fuels (e.g., crop biomass such as bagasse), the above assumptions and conditions are not applicable. The L value is much more fuel composition and system efficiency dependent. Capturing BFC energy credit for the use of biomass fuel in place of fossil fuel (e.g., co-generation, pre-heating a process stream) requires consideration of system application specifics.

BFC Models

The following application of the BFCM to energy and environmental scenario models uses representative as opposed to all inclusive literature data. The purpose is to illustrate the use of the methodology for a few BFC data sets. In the present treatment, the parameters of interest are specified using British thermal unit (Btu), Acre, Gallon (Gal), and Bushel (Bu) units.

Analyzing Yield Aspects

The two main BFC liquid biofuels products are ethanol (e) and biodiesel (d). Consider the created ethanol fuel energy per acre for the corn to ethanol BFC where the portion F of corn processed through the wet versus dry milling is varied. Based on equation 10.1 the energy-yield relationship is:

Ee/A(Btu/Acre) = YC [YD F + YW (1 ″ F)]Ebiofuel e

Here YD is the Ybfp for corn to ethanol Dry mill processing, YW is the Ybfp for corn to ethanol Wet mill processing, F is the fraction of ethanol corn Dry mill processed, and Ebiofuel e is the ethanol UE fuel value.

From a local/regional and national perspective, the potential gain from BFC improvement is an important consideration. The equation 10.1 Ee/A yield relationship provides insight into such considerations. Large variations in corn yields occur as the result of soil, weather, and crop management practices: 85–245 Bu/Acre (Dobermann and Shapiro, 2004).

For biorefinery yields in the 2.6 Gal/Bu range, a region producing at 140 Bu/Acre will attain Ee/A values 25% higher than a region producing 112 Bu/Acre. Alternatively, processing the 112 Bu/Acre region corn at a 2.8 Gal/Bu biorefinery achieves 8% higher Ee/A value over the 2.6Gal/Bu facility. A subset of this is Wet versus Dry mill utilization considerations illustrated in For the soybean to biodiesel BFC the created biodiesel energy per acre is:

Ed/A(Btu/Acre) = YS Yd Ebiofuel d

Combining the corn and soybean crop rotation and fuel production BFC's:

Eed/A(Btu/Acre) = YC CR [YD F + YW (1 " F)] Efuel e + YS (1 " CR)YdEfuel d

Here Eed/A is the combined energy content of ethanol and biodiesel fuel produced and CR is the crop rotation cycle fraction for corn planting (e.g., alternating plantings:

CR = 0.5; 2 out of every 3 plantings: CR = 0.67).

The CR needed to achieve an equal energy gain from each crop in the cornsoybean BFC is given by the relationship:

CR = YSYdEfuel d/[YCYMillEfuel e + YSYdEfuel d]

Here [YD P+YW (1"P)] is defined as the corn to ethanol effective processing yield YMill. To achieve parity under the 'current yields' requires a 5 plantings crop rotation sequence comprised of 1 corn planting for every 4 soybean plantings. The alternate year crop rotation sequence approaches parity for the low corn and high soybean yields. Again the analysis does not include NEB aspects.

OUR FOOD AND FUEL FUTURE

During the past century, inexpensive fuels and an outpouring of new science and resultant technology have facilitated rapid growth and maintenance of human populations, infrastructures, and transportation. Developed countries are critically dependent on the liquid fuels required by present day transportation of goods and services and by agriculture and are dependent on various fuels for generation of electricity. Authorities and the media present physical growth as an economic and social need, but consumption and its growth ultimately cause declining availability and increasing price of fuels and energy. Increased burning of carbon fuels with increase of carbon dioxide in Earth's atmosphere is the principal cause of increasing global warming, which is well-measured and a probable source of future disruption of world ecosystems.

Regrettably for humanity, the power of new technologies has not yet been accompanied by vitally needed political and cultural developments in the U.S. and in many other countries. The political system in the U.S. seems unable to mitigate processes that contribute to global warming nor adequately address declining supplies of liquid fuels, nor does it discourage social pressures for continued physical growth. Search for alternative sources of liquid fuels for the transportation sector in developed countries and in the United States in particular produce strong connections among energy supply, food supply, and global warming. Various current U.S. programs are examined and none appear effective toward prevention of a future disaster in human terms. The social organism is not ready now to sacrifice for future gain or even for sustainability.

Connections among energy supply, food supply, global warming, and political campaigns have become strong in the United States during first years of the 21st century. Liquid fuels derived from petroleum are of enormous importance in developed countries because they are a principal support of the transportation industry (and petroleum- and coal-derived hydrocarbons are also critical ingredients in the chemical industry). Demand for liquid fuels continues to increase, but discoveries are tapering off, and sharply increased price is stimulating search in the U.S. and other nations for sources other than the traditional oil industry, which involves a dependence on foreign suppliers of uncertain reliability. The search for suitable alternatives is influenced and befuddled by powerful established interests whose primary goals are their own economic benefits rather than societal welfare. Several of the programs are examined in detail in following pages, and it should be borne in mind that numerous proposals reflect wishes of special interests more than conclusions from rational analysis. Controversy abounds.

PRICE AND AVAILABILITY OF TRADITIONAL FUELS

Traditional energy sources, i.e., those that produce a substantial amount of the power currently used, include coal, oil, natural gas, hydropower, and nuclear fission. Non-traditional sources, i.e., emerging sources, some on trial or subjects of significant experiments, include wind, tides and river currents, solar, hydrogen, biomass, geothermal, and nuclear fusion. Brief comments on all of these energy sources follow, with much of the presented data obtained from the U.S. Energy Information Administration.

COAL

Coal burning produces about half of all the electrical energy1 produced in the United States, a ratio that has remained nearly constant for the past twenty-five years, even as electricity usage has increased 70%. Coal is usually said to be so abundant in the United States that its use as an energy source here will endure for centuries. Next to hydropower, it is the cheapest source of energy, and about 85% of the 1.1 billion tons produced and consumed annually in the United States is bituminous coal and is used within the country to generate electricity.

Coal burning in the U.S. produces annually about 2.1 billion metric tons of carbon dioxide,2 the major contributor to global warming. The carbon dioxide is emitted to the atmosphere and it is buried permanently (sequestered) only in rare situations where, under high pressure, it enhances tertiary recovery of petroleum. Coal burning has increased 19% since 1990 but was down nearly 1% between 2005 and 2006 because the average U.S. winter in 2006 was milder and the summer cooler than in 2005.

According to EIA data, the price of coal as delivered to power plants in the United States is significantly variable with region, costing much more in New England (<"$65/ton in 2005), for example, than in the Midwest (<"$20/ton), and owing to increasing world demand, the price is rising as this chapter is developed. In 1975 there was a temporarily doubled price that was largely caused by the Arab oil embargo of 1973, and this peak was followed by a slow decline of coal price. An important way of looking at the price of coal is through energy content – a typical minehead price in 2005 was about $1.15 per million BTU, or about $20/ton for coal with a 50% carbon content and the delivered price was about $45/ton, but variable depending on the distance from mine to user.

Past sulfurous emissions from coal-burning power plants have been widely associated with "acid rain", which causes corrosion and has altered the pH and ecology of some lakes, especially in northeast U.S. The Shady Point power plant at Panama, Oklahoma, which started in commercial operation in 1991, avoids sulfurous emissions by mixing local high-sulfur coal with limestone, also mined locally. As the limestone is heated, it emits carbon dioxide and combines with the sulfur, producing calcium sulfate, which in another form is known as gypsum. Some of the slag finds a use in neutralizing pollution and some finds use as a road stabilizer, though most goes to land-fill sites.

The Shady Point power plant produces its maximum 320 megawatts throughout 24-hours during June-August while burning daily about 3000 tons of Oklahoma coal mixed with about 1000 tons of limestone. The average sulfur content of the coal is about 3% and its carbon content is variable from about 55% to 70%, depending on mine origin. Its carbon dioxide emissions during summer, based on 60% carbon in the coal, are thus about seven thousand tons daily with about 6% of that from the limestone, and 200 tons/day are extracted from the flue gas as food-grade CO2. The augmentation of CO2 by limestone seems unimportant in view of the large ongoing emissions from other coal-burning power plants. (Personally communicated, 2007; also see Shady Point website).

Most actual reductions of sulfur emissions in the U.S. have resulted from use of low-sulfur coal from Wyoming instead of coals with higher sulfur content from Oklahoma and eastern U.S. Particulate emissions from coal-burning power plants, another cause of "acid rain", have also been greatly reduced in recent years. Emissions from coal burning include mercury and other heavy metals including arsenic, uranium, and thorium. During 1999–2003, the U.S. Environmental Protection Agency collected and analyzed fish tissue from 500 ponds and lakes across the United States for a wide range of elements and organic toxic chemicals.

Levels of mercury or arsenic exceeding EPA screening levels for human health were found in many of them. This contamination is attributed to coal burning, though it seems that this attribution has not been proved. Of twenty one sites sampled in Oklahoma, nine had levels of mercury or arsenic that exceeded EPA screening levels3 (Environmental Protection Agency, 2007), and many states have issued directives concerning permissible limits on eating fish so contaminated.

Questions have been raised about prospects for the enduring use of coal owing to environmental concerns, possible exaggeration of reserves amenable to economical extraction, and probable increased future costs of transportation (Schneider, 2007a).

Further concerning the environment, coal mining in the U.S. state of West Virginia has become very controversial because whole mountain tops have been moved into adjacent valleys in order to expose coal seams. This has caused marked deterioration of water quality and other environmental abominations. Mine safety also continues as a major issue with strident public calls for additional regulation by the U.S. federal government.

China and the United States in 2007 emit nearly equal amounts of carbon dioxide, and further major development of the coal industry in China's Shanxi Province was outlined in a special supplement to *China Daily*, published September 18, 2007. Substantially increased production of raw coal, liquid fuels from coal (usually, Fischer-Tropsch process), and coalbed methane (see following section) were projected during the Taiyuan4 International Coal and Energy New Industry Expo 2007. This development is seen in China as essential to improved prosperity of the country and its people.

The indicated environmental negatives diminish as advanced technologies are applied. Coal combustion seems destined to remain for decades as a major source of electrical power. However, in spite of promulgation of State policies toward energy conservation and emission controls such as presented by the Shanxi Minister of Commerce, serious concerns persist because coal burning and coal conversion are major producers of carbon dioxide, the principal contributor to global warming.

Natural Gas

At the start of the 20th century, natural gas was a little-desired byproduct of the petroleum industry and sold for as little as five cents per thousand cubic feet at the wellhead. During the 1970s, price rose from 17 cents to $1.20 per thousand cubic feet, and during the 1980s and 1990s, natural gas was irregularly priced, but sometimes above $2.50. A substantial price rise to 2007 levels fluctuating between $5 and $7 per thousand cubic feet began about the year 2000. Improved technologies of horizontal drilling and fracturing in tight rock formations have enabled gas production in areas of shale and coal formations in the United States, and the high cost of production is supported by high price of the product. Regrettably, modern methods of extraction often degrade soil and water.

Natural gas is widely used today for home heating and for standby power generation, and gas-to-liquids technologies are being proposed for production of liquid fuels. Gas production and consumption in the United States has been nearly steady at about 24 trillion cubic feet annually since the mid-1990s, and challenges to maintain that level of usage in the presence of an ultimate decline of U.S. supplies have led to proposals for importation of liquefied (strongly cooled) gas (LNG) from the Middle East. However, proposed LNG terminals are often opposed by local groups apprehensive of explosion dangers.

Natural gas is also used for production of the fertilizer bases, ammonium nitrate and urea. As the price of natural gas has risen, its preferred use for home heating and power generating facilities has led to closure of about 40% of U.S. fertilizer production capacity since 1999 and to increasing importation of nitrogen fertilizer from regions where natural gas is much less costly than in the U.S. Imports now account for a little more than half of total U.S. nitrogen supply, which has remained nearly steady at twenty million product tons since 1998. A recently developed controversy within the United States involves proposed new facilities for electric power generation, with natural gas interests pointing to the lower carbon dioxide emissions associated with natural gas, and coal advocates indicating lower costs with coal.5 In any case, creation of new power plants, whether gas- or coal-powered, to accommodate continued physical growth leads to increased CO2 emissions and exacerbation of the global warming phenomenon.

It is conceivable that further research will lead to a vast expansion of natural gas supplies and, perhaps, to a medium for the more effective storage of hydrogen than is available today. Such advances could involve clathrate hydrates, which are abundant below permafrost and along continental margins in and beneath waters whose temperatures are near water's freezing point. Clathrate hydrates are solid combinations of hydrocarbons, especially methane, or carbon dioxide with water. It is estimated that several times the known traditional resources of natural gas are so combined, and there is concern that global warming will lead to release to the atmosphere of vast quantities of clathrate methane. This would be especially important because methane is about 20 times the greenhouse gas that is carbon dioxide. While many clathrate deposits have been identified, an effective technology for methane extraction has not been developed. Mao, et al. (2007) describe the situation in desirable detail, and their article contains a substantial list of references.

PETROLEUM

A direct use of oil is for home heating, especially in northeastern United States, and oil refined to gasoline and diesel fuel provides more than 95% of the energy used in the U.S. transportation industry. Oil production in the U.S. peaked at 9.5 million barrels per day in 1970, in close agreement with a prediction of M. King Hubbert.6 Since 1985, U.S. crude oil production has declined every year, and in 2005 was 5.2 million barrels per day. And, as a result of both declining domestic production and increasing demand, crude oil imported to the United States increased from 5.8 million barrels per day in 1991 to 10.1 million barrels per day in 20057. Total U.S. consumption of crude oil and other imported petroleum products continues to rise about 1% annually, and totaled 20.8 million barrels per day in 2005.

In the early 1970s, the inflation adjusted price hovered near $10/barrel, but it is near $90 and rising irregularly as this article is completed at the end of October 2007. The price of crude oil is reflected in the price of refined products, and gasoline in June 2007 cost as much as $4/gallon in some U.S. markets, and more than $3/gallon on average nationwide.8 Dependence of the U.S. for oil from foreign sources of uncertain reliability, rising prices, and concern for competition and projected future scarcity (e.g., Simmons, 20059; Ghazvinian, 2007) are stimulating search for alternative motor fuels, discussed further below. But a major concern arises because all carbonaceous fuels produce carbon dioxide emissions that contribute to global warming, and emissions by the U.S. transportation sector are about one third of the total.

A striking example of conflict between efforts to gain access to new oil and the greenhouse problem is provided by the tar sands of northern Alberta. Economically recoverable reserves of heavy oil there are estimated to well exceed one hundred billion barrels, which would supply the whole world for several years at the present rate of consumption (about 30 billion barrels annually).

But the extraction process is very energy intensive, involving mining of the sands, their transport in huge trucks to crushing and heating facilities, and costly refinement and transport of a still tarry product via pipelines. In situ heating with large use of water is also implemented for recovery of oils at depth. These energy intensive processes produce much greater release of carbon dioxide than is released during recovery of lighter oils by traditional methods. The processes for recovery of tarry oil are described at length in a supplement to *E&P Oil and Gas Investor* (Hart Energy Publishing, 2006), which includes a list of companies and their plans to invest $80 billion in Alberta oil sands by the year 2014. Discussion of advanced technologies for extraction and refinement of tarry oil has also been presented (Hart Energy Publishing, 2007).

Hydropower

Most dams are built for flood control and irrigation, but hydropower provides about 7% of all the electricity produced in the United States. The largest hydroelectric facility in the U.S., Grand Coulee Dam, serves multipurposes while providing average power of about 2300 megawatts, the equivalent of two or three ordinary coalburning plants. In the U.S., it is not expected that additional hydropower can be provided in quantity sufficient to replace other energy shortfalls, but in China, the Three Gorges Dam is scheduled for completion about 2010 and should provide 18 thousand megawatts of electricity.

Dams do have negative effects. Thus, sediment tends to accumulate behind dams, reduced sediment in downstream flows usually fails to compensate for erosion of river deltas, and there are often adverse effects on fisheries.11 For such reasons and others, especially the destruction of agricultural areas flooded by impounded waters, the construction of hydroelectric facilities produces controversy, and some existing dams have even been proposed for removal.

Nuclear Fission

Studies in astrophysics and atomic physics subsequent to presentation of Einstein's special and general theories of relativity in 1905 and 1916 showed paths for producing enormous energies by conversion from matter. Heavy elements, including uranium, are produced during the collapse of stars much more massive than Sun, and the products of the radioactive decay or fission of the heavy elements are less massive than their sources. The mass difference appears as energy. Uranium is widely present on Earth, its average concentration is near three parts per million, and it is over ten times more abundant than silver, for example. It consists mainly of the isotope 238U, with about 0.7% 235U, which is principal reactor fuel. For purposes of power generation 235U is concentrated to about 3% by an energy-intensive gaseous-diffusion process that takes advantage of the slight difference of atomic weights among isotopes. During typical reactor operation, atoms of 235U absorb neutrons and then split into other elements with release of energy and neutrons. The reaction is initiated by stray neutrons and maintained by those released. Materials that absorb neutrons are arranged to maintain a concentration of neutrons that produce heat at the desired rate. The energy statistics are startling: Fission of one kilogram of 235U produces as much energy as combustion of about 40 million kilograms of TNT and without any greenhouse gases.

As in other power plants, the heat generated by controlled fission is used to boil water and create steam that drives turbines to generate electricity. At this writing, nuclear fission provides about 19% of all electricity in the U.S., 16% worldwide, 30% in Japan, and maximally 78% in France. According to the U.S. Energy Information Agency, there were 436 operating reactors in 30 countries worldwide during May 2007, including 103 operating reactors in the United States. There is little question that nuclear reactors could provide abundant electricity but their future is clouded by risk of accidents that degrade wide areas, such as occurred at Chernobyl, by risks from terrorism, and by risks attendant to disposal of highly radioactive nuclear waste for hundreds of thousands of years. Possible effects of seismicity and volcanism at the proposed U.S. disposal site at Yucca Mountain, Nevada, have been examined by Hinze, et al. (2008).

And use of breeder reactors, so-called, which convert uranium of molecular weight 238 to fissionable plutonium of weight 239 and could provide a nearly endless energy supply, is inhibited by fears that the process of separating plutonium from the mix would be adapted to bomb making. Although more than thirty new nuclear plants are under construction in twelve countries as this chapter is prepared, new construction in the United States has been strongly inhibited by negative public opinion. However, the combination of conditions described in preceding sections, coupled with reactor designs that are much improved with respect to simplicity and safety may well lead to a resurgence of fission reactor construction in the U.S.

In this matter, a paper on net energy (Tyner12 2002), should be examined. Owing to energy requirements for construction, operation, waste disposal, and ultimate dismantling of nuclear power plants, Tyner concludes, "any expectation that Nuclear Power will be a viable substitute for fossil fuels is, at best, questionable". There is also the matter of carbon dioxide releases that attend manufacture of the cement and steel needed for reactor construction and the mining and refinement of nuclear fuel. Details are complex and this author proposes that the matter of net consequences be carefully examined. In any event, while electric power however generated is a poor direct substitute for liquid fuel for transportation in 2007, electrical energy can be used for the manufacture of liquid fuels.

Alternative Sources of Energy

As already noted, the high and rising price of oil and its derivative fuels is a principal accelerant to search for alternative fuels. Another motivation for this search lies in concerns about global warming, produced by increasing emissions of carbon dioxide during transportation, power generation and during manufacturing processes attendant to production of steel and cement, for examples. As shown below, it will be difficult to develop an alternative fuel pathway that supports either generation of electricity without excessive carbon dioxide emissions or an automotive industry with markedly reduced usage of petroleum and its products. Further, the programs so far implemented in the United States appear to be means for accumulation of wealth by a relatively small number of beneficiaries who have both the power to control legislation and ability to create a public perception that realistic steps are being taken when the fact is opposite. The incorrect public perception allows business to proceed as usual even though collapse may be just around the corner. We first discuss several suggested alternate energy sources that may be contributing in a small way, and then we consider possibilities whose successful future application must depend on research results so-far elusive. Then we take up nationally empowered programs involving biologically based fuels.

Wind, Rivers, and Tides

Wind has been used for thousands of years for sailing and for grinding grains, and decades ago in the United States there were, beyond the range of utility lines, many small windmills that powered a few light bulbs and radios. Small windmills are still widely used in western United States to pump water for livestock.

Modern wind energy units are especially valuable in remote communities where electricity is otherwise supplied by small diesel-fueled installations, which can be very costly. According to the Energy Information Administration, wind began to be a significant source of electricity in the United States about 1990. Wind power technology has advanced steadily and large machines now deliver up to five megawatts each during favorable winds. Use of wind power has advanced with particular rapidity in Europe, and Denmark, an acknowledged global leader in wind energy, derives approximately 20% of its electricity from wind turbines and plans for an increase to 50% in 2030. The increase in wind energy production since about 1980 in Denmark has enabled that country to stabilize its carbon dioxide emissions.

Technological advances have greatly reduced the price of power from wind, and land-based wind turbines now cost from $1500 to $3000 per kilowatt, nearly competitive with coal-burning power plants. According to the American Wind Energy Association (2007), the most efficient wind generators in windy places can deliver power at a cost of five to ten cents per kilowatt hour. This is similar to the charge imposed by most utilities in the U.S., but wind power in the U.S. is still subsidized with a federal tax credit of 1.5 cents per kWh.

Electricity is produced by wind with no gaseous emissions at all, though emissions occur during manufacture of the steel, concrete, and other items used in fabrication and erection of the turbines. Where winds are favorable, the overall payback is large, however, and is still increasing with technological advances. The great height, several hundred feet, of modern machines places them above the layer where friction with the ground causes a strong diurnal variation of wind – at the greater height the average wind is nearly constant throughout the average day. Since the rate of electrical power generation is proportional to the cube of the wind speed, site selection is very important. Site selection in Oklahoma has been aided by a network of over one hundred weather-reporting stations within the State (Kessler, 2000; Oklahoma Mesonet, 2007).

The capacity of electricity production from wind is increasing in the U.S., with approximately 5000 megawatts added during the two-year period 2004–05. Subsequent additions brought the total U.S. wind power capacity to 12,634 megawatts as of June 30, 2007, more than one percent of the U.S. total of about one million megawatts (See footnote 2). Production of electricity from wind does seem to be a good, but, as noted elsewhere (e.g., Tyner, 2002), ". . . even if wind machines were constructed everywhere it is practical to erect wind machines in the United States they would only be able to provide a pitifully small fraction of the net energy compared to that needed to power the industrial economy of the United States. . . " This seems true in Oklahoma, although five wind farms have been installed and others are planned. Installed wind capacity in Oklahoma totaled 690 megawatts in August, 2007, about three percent of Oklahoma's electric generating capacity (American Wind Energy Association, 2007; Oklahoma Wind Power Initiative, 2007).

Capacity and capacity factors can be confusing. Because wind is highly variable, the average generation by a wind farm is almost always less than half of its capacity with optimum wind, and one third is often taken as a standard. This means that Oklahoma wind farms can presently provide, on average, about 1% of the power that can be provided by traditional facilities. Furthermore, since electricity cannot be economically stored,15 no amount of wind power installation allows reduction of the number of power plants fueled by coal, natural gas, or nuclear fission, except to the extent that consumers agree to interruptible power supply. Of course, during windy periods, power generators that use fossil fuels can be cut back, thereby reducing emissions and saving non-renewable fuels.

At this writing, wind farms have been proposed offshore Cape Cod, Massachusetts, and offshore south Texas in the United States, but are attended with uncertainties in both costs and esthetics. Research at the Massachusetts Institute of Technology (MIT) envisages anchoring systems for wind farms offshore that would withstand the force of wind and wave in hurricanes at a distance beyond objections from onshore landowners (Anthony, 2007). Average wind at sea is much stronger than on land, and power generation offshore could reverse Tyner's findings. Associated costs and other results of this research remain to be seen. Utilization of river and tidal flows for energy generation is closely related to wind power technology. Some experiments in Europe were undertaken forty years ago, and there is more activity today, both in Europe and North America. Newspapers have discussed additions of turbines to an experiment ongoing in the East River, New York, and there are proposals for major installations in San Francisco Bay and elsewhere. The sea and rivers harbor enormous energies in waves and flows, but practical utilization is very challenging. Further experiments with river and tidal flows will probably be encouraged and developed with reasonable government assistance.

Solar Power

The diameter, D, of Earth is 12,750 kilometer, and its cross-section is ðD2/4 = 1.28 × 1014 square meters. Solar radiation on a flat plate perpendicular to the rays outside Earth's atmosphere is 1.4 kilowatts per square meter.16 Thus, Earth intercepts 1.8 × 1017 watts of solar energy, i.e., 1.8 × 105 terawatts, which is about fourteen thousand times the rate at which humankind produces energy from a combination of fossil fuels, nuclear, hydropower, and wood and other biomass.

Use of solar energy is prima facie attractive because there is so much of it and because its use has little environmental impact. It may be used in two distinct ways: conversion to electricity and direct heat. The former is presently about ten times more costly than production of electricity by traditional means. An average of ten percent of U.S. electricity would be produced from solar panels of ten per cent efficiency on sunny days from an area of about 180 square kilometers (67 square miles). While this is a very small fraction of Earth's surface, it is a large area in human terms. Power generation would be maximum during the day and zero at night, and unless means were provided for storing produced power and distributing it to meet variable demand, it would be a back-up facility on sunny days to reduce demand for power generated by other means.

The energy and research sides of conversion of solar radiation to electricity are well discussed and explained in *Physics Today* (Crabtree and Lewis, 2007) and, with other energy discussion, in

Science. Current research and development suggest that efficiencies for conversion of solar radiation to electricity may be doubled within a few years. Even with low conversion efficiencies, communication is much enabled today with panels that produce a few tens of watts for radio links in many field applications without need for connections to a utility's grid, and small solar electric units at reasonable prices maintain electric fences on farms and ranches where access to utility lines is not easily available.

As a direct source of heat, solar radiation does have important practical applications today in water heating, and the design of solar collectors for that purpose has been recently improved with vacuum components manufactured in China (Apricus. com, 2007). Solar water heaters allow avoidance of use of electrical energy for heating, but in cold climates some regrettable complexity is needed in the form of heat exchangers to prevent damage incident to freezing. Solar cookers can be quite effective when Sun is high and skies are clear; your author enjoyed such for several years at his home on an Oklahoma farm and saw several in use in a monastery during a trip to Tibet.

Major solar installations of both the photovoltaic and direct heat types are on line in California and Nevada, USA. For direct heat, known as concentrated solar power (CSP), hundreds of mirrors track Sun and reflect its energy to a tower where the concentrated solar radiation flashes water to pressurized steam at 250C for driving turbines. Another direct heat technology, uses a series of parabolic troughs that focus Sun's energy on a central pipe and thereby heat oil therein to about 400C. The oil flows to a steam generator connected to a turbine for generation of electricity. A new CSP facility is currently under construction near Las Vegas, Nevada and a photovoltaic facility is expected to be on line at the end of 2008 with fourteen megawatts for Nellis Air Force Base, also near Las Vegas. Use of solar direct heat is being realized in experimental new power plants in Spain and in Algeria (Trade Commission of Spain, 2007). The two methods noted above are subjects of major experiments by a subsidiary of Abengoa, a holding company. A heat storage mechanism involving troughs 18-feet wide with 28 thousand tons of liquid salt is also being developed in Spain. Planned for completion in 2012, the so-called

Sanl´ucar La Mayor Solar Platform should generate more than 300 megawatts of solar power with both of these technologies and photo-voltaic panels as well.

The government of Algeria plans to invest in solar power some of its revenues gained from exports of oil and natural gas, about $55 billion annually at this writing. The firm, New Energy Algeria, established in 2002 to exploit renewable resources, has partnered with Abengoa for construction of a 150 megawatt power plant that combines the solar resource abundant in the Sahara desert with generation of electricity by natural gas. It is reported that the company hopes to produce six thousand megawatt capability by the year 2020 and export that to Europe via cables under the Mediterranean Sea. The first Algeria facility is projected to use cogeneration with natural gas to fill gaps at night and during occasional cloudy periods.

Hydrogen and Batteries

Numerous research challenges and prospects for a U.S. hydrogen economy have been detailed by Crabtree, et al. (2004), and widely discussed by media. It is not expected that hydrogen would be used directly as an automotive fuel because pure hydrogen is very difficult to store in quantity. But use of hydrogen is attractive because the product of hydrogen oxidation in fuel cells is simply water, and there is no attendant environmental contamination. Perhaps the most important of present applications of hydrogen as a fuel are in the U.S. space program, and there are automotive trials in a fuel cell program that is highly experimental. The fuel cell is properly regarded as an energy storage device, as is a battery. Basic to development of a hydrogen economy would be economical means for production of hydrogen in much larger amounts than produced in the present chemical sector of the U.S. economy. Hydrogen is almost ubiquitous but is tightly bound in water and other substances. In addition to the research that would be essential to development of acceptably economic means for hydrogen production, infrastructures for storage and transport of hydrogen would have to be created.

The amount of energy used for hydrogen production is several times the energy of the hydrogen produced. Partial justification for expansion of a hydrogen production industry might be found in the burning of abundant low-cost coal as a source of the electrical energy needed for hydrogen production by disassociation of water, but greenhouse gas emissions with coal burning are inhibiting. Of course nuclear power could also be used, but expansion of the nuclear industry is inhibited by concerns for contamination and disposal of nuclear waste. Expanded use of solar power may represent an ultimate good source of energy for hydrogen production. The challenges for hydrogen lie in development of economies in all of production, storage, and distribution, and numerous research efforts are underway. If batteries could be developed to the point that they would safely and economically provide the range, power and rapid "plug in" recharge that automobile users want from their automobiles, there could be significant savings of liquid fuels. Batteries used in laptop computers during the year 2007 have very high energy densities but have had safety problems. If safety were assured along with achievement of economic gains through further research and large scale production, electric automobiles powered by numerous laptop batteries could become a reality, as discussed by Schneider (2007b). Further background is available on numerous web sites.

GEOTHERMAL

Earth's interior heat has been used for human needs for thousands of years. Hot springs have been used for baths, and today in Iceland, a volcanic area, geothermal sources provide 40% of Reykjavik's hot water! In addition, there are about 20 hectares of geothermally heated greenhouses in Iceland for production of fruit, flowers, and vegetables. However, expansion of greenhouse production in Iceland is inhibited by low levels of natural illumination, which leads to implementation of artificial lighting. More important, Iceland's self-sufficiency is presently impeded by the availability of lower-priced imports, which provide about 75% of Iceland's fruits and vegetables.

Use of geothermal heat for electric power generation dates from 1904 at Lardarello, Italy, where local volcanism provides heat sources near Earth's surface. In the United States, some twenty power plants at the Geysers, north of San Francisco, California, provide 850 megawatts of power from dry steam – this comes from strata less than three thousand meters below the surface, and the total amount of electrical energy produced is similar to that provided by one typical coal-burning facility. MIT professor Jefferson Tester recently noted that Earth's interior heat, if accessed much more widely for power generation, could provide humankind's demand for power generation for thousands of years (Bullis, 2006). And Roach (1998) has noted that about 99% of Earth's total mass is at temperatures between 1000 and 5000C. However, the necessary heat must be found in a thin surface layer within which the average rise of temperature with depth is about 25C/km. Temperatures near 200C are necessary for viable power generation from geothermal heat, and, owing to spatial variations in the rate of temperature rise with depth, there are many places where wells to depths of about five km find the desired temperatures. Possible applications of geothermal heat are becoming more promising owing to major advances in the drilling technologies applied to recovery of oil and natural gas, particularly in the technologies of horizontal drilling and rock fracturing.

An important geothermal experiment ongoing at this writing near Basel, Switzerland, illustrates both potential and pitfalls (H¨aring, et al. 2007). In addition to a field of monitoring wells, three principal wells for the facility were planned initially in Basel, one for water injection and two for production of hot water. It was planned to deliver about 3.5 megawatts of electrical power to the grid and the equivalent of about 5.5 megawatts of heat for local heating. However, initial tests were accompanied by earth tremors sufficient to produce significant apprehension in the local population and a flurry of claims for minor damage, and at this writing (September 2007) the project has been stopped pending further assessments. As this is written, only about 1500 megawatts of electricity is provided globally from geothermal sources – this is comparable to the production of one large coalburning plant or two ordinary facilities.

NUCLEAR FUSION

Fusion, in contrast to fission, involves combination of light elements to make more massive elements whose atoms weigh less than the sum of those used for their creation. As with creation of the fission element, uranium, this is a process that takes place in massive stars. Under extreme conditions of temperature and pressure, light elements beginning with hydrogen are fused into heavier elements, ending with collapse of the star and creation of elements heavier than iron, including uranium and some highly radioactive transuranic elements. Elements lighter than iron produce energy when fused; heavier elements produce energy when split.

Hydrogen, consisting of one proton and one electron, constitutes about 74% by mass of the known universe, and most of the balance consists of helium, with only about 2% represented by all other elements. On Earth, hydrogen is about 11% of the mass of the oceans, with deuterium (hydrogen of mass 2) comprising about 1/70% by mass of the total hydrogen. A third isotope of hydrogen, tritium, with two neutrons and one proton, is of importance because of a prospect of its use in a fusion process that may someday be perfected on Earth.

While energy production by fission of uranium is well-established world-wide, energy production by fusion of hydrogen, akin to a controlled hydrogen bomb, is still in its infancy and may never be feasible on Earth. However, effective fusion technology is much sought because it would produce no long-lived radioactive aftermath nor carbon dioxide, and does not, per se, have implications for nuclear war. And centrally important, if the technology for energy production via fusion were perfected, the production of electricity sufficient for any purpose of humankind could be limited only by the number and power of fusion reactors constructed. Recent history and technical challenges facing the international fusion program have been presented in *Science* (Clery, 2006). The effort toward power by fusion began in several countries during the 1950s. In 1985, programs in separate countries began to be internationalized after a summit conference at Geneva produced agreements between Russian premier Gorbachev and U.S. President Reagan. The program is known as ITER – International Thermonuclear Experimental Reactor. Its latest manifestation is an agreement among seven governments17 to construct an experimental reactor in Cadarache, in southern France, at a cost presently estimated near $12 billion over ten years. After construction, the facility would be run for twenty years to develop improved knowledge of a proper subsequent design. It will be enormous and very unlike any existing power plant on Earth.

It is presently believed, on the basis of numerous ongoing experiments, that this greatly scaled-up facility will demonstrate net generation of power, but the technical challenges are awesome. Basically, the problem is to replicate on Earth the very high pressure and high temperature conditions in stellar interiors. This would be accomplished with strong electric currents that produce a strong magnetic force and a pinch effect.18 The zone of extreme temperature must be held away from the walls of the facility because contact would reduce temperature by conduction, the magnetic fields must be controlled to prevent instabilities in the toroidal active zone and the materials used must resist embrittlement by radiation.

ITER fuel consists of a mixture of deuterium and tritium, the former separated from water by distillation and the latter produced in the reactor itself. At sufficient temperature and pressure the velocity of the hydrogen atoms becomes large enough to overcome the electrostatic repulsion of the nuclear protons, and helium and energetic neutrons are created.

ITER construction is scheduled to begin in 2008, and orders being placed at this writing include such costly items as superconducting magnets. The outcome is uncertain, but potential reward is enormous, and "nothing ventured, nothing gained". Electricity satisfies many needs and can provide the energy needed for manufacture of liquid fuels.

BIOFUEL RESEARCH, ETHANOL AND BIODIESEL

Search for a biological base to alternative fuels is wide-ranging. In 2007, the U.S. Department of Energy provided $375 million over five years to establish bioenergy research centers at the Lawrence Berkeley National Laboratory in California, the University ofWisconsin at Madison, and at Oak Ridge National Laboratory. Efforts at these centers will be focused on devising biological processes to convert cellulose to liquid fuel. The research presumes that success could be followed by viable harvesting of cellulosic materials of forest products, grasses, and crop residues, but as mentioned again in the last paragraph of the next section, impacts on agricultural practice and land use may be unsustainable.

In related research at the J. Craig Venter Institute in Rockville, Maryland, some studies are focused on creating bacteria that contain the genomes for making biofuels from cellulose (Pennisi, 2007).

Whether or not research such as described in the preceding two paragraphs is "successful", both it and its possible future applications will assuredly be controversial. Humankind already consumes a large fraction of the energy represented in annual biological growth,19 and our search seems directed toward new modes of exploitation rather than toward carefully planned elimination of waste and reduction of demands on non-renewable resources. The ethanol and biodiesel programs described in the following two sections, except for conceptual production of ethanol from cellulose, use already developed technology for production of liquid fuels from the biosphere.

Ethanol from Corn, Sugar, and Cellulose

Much of the following discussion is well presaged in a pamphlet distributed twentyseven years ago from the Federal Reserve Bank of Kansas City (Duncan and Webb, 1980). The FRB report appears to have been prompted by concerns arising from the embargo placed on export of Arab oil to the United States in the 1970s. Concerns with prospective declines of petroleum-based gasoline also led to a more formal examination of conversion of biomass to ethanol (Energy Research Advisory Board, 1980 & 1981). Despite the substantial negative energy conversion ratio presented by these reports, interest in ethanol production as a substitute for gasoline has increased and, in the United States, has culminated in Congressional legislation which calls for production of 36 billion gallons of biofuels by 2022. But will this be achieved, and should it be achieved? At this writing in mid-2007, production of ethanol from corn in the United States is at a rate of about six billion gallons annually, having increased from one billion in 1990. The United States among nations thus leads annual production of ethanol, having recently replaced Brazil.

Ethanol from corn is produced by first mixing finely ground corn with water and adding enzymes alpha amylase and glucoamylase to the warmed mixture for conversion of the starch to glucose. Ethanol is then produced from this simple sugar by fermentation with yeast, and the ethanol is concentrated by distillation.Well over one hundred ethanol plants have been built during the past few years at a cost of more than $50 million each in the United States and several tens more are planned. A typical plant consumes about fifty thousand bushels of corn daily, 20 million bushels annually, and produces about one million barrels (@ 42 gallons) of ethanol annually, which is roughly equivalent to 5% of U.S. oil consumption for one day. The total investment in ethanol plants is thus about $6 billion and the yield of six billion gallons of ethanol is equivalent to 4.5 billion gallons of gasoline, equivalent to five or six days supply of oil in the United States.

Every day, the public is swamped by media presentations pro and con, reflecting intense controversy. There are a host of arguments against this program, and, in your author's opinion, this program and several others have gone forward either because lobbyists effectively buy legislation with contributions to legislators or instill fear among candidates that elections will be lost if programs desired by special powerful interests are not supported. The ethanol program will ultimately prove to be destructive. Consider the following.

First, the net energy argument concerning corn-to-ethanol: Prominent contradictory analyses have been presented (Pimentel, et al., 2007, and Shapouri, et al., 2002). The former, in agreement with earlier studies, finds that more energy is required to grow, harvest, and transport corn, ferment it to ethanol, and distill the ethanol to increase its purity to 90% or more, than is obtained from the ethanol. The latter finds the opposite to be true. The ratios of input to output energies presented by the two studies are within the limits 1.5:1 and 0.5:1. It is important to note that Pimentel, et al., includes some energy inputs that are admittedly omitted in the analysis by Shapouri, et al. It is critically important to observe that western societies cannot function in the manner to which they have become accustomed with either ratio, because they are drastically unfavorable with respect to historic oil, said to have been 0.01:1 during early days of discovery and exploitation, and increased to about 0.05:1 today, owing to high costs of recovery in hostile environments.

Further in connection with the energy ratios, recall that any conversion process involves energy loss. For example, the energy in gasoline is less than that in the oil from which it is refined. But we make gasoline from oil because gasoline has higher uses than crude oil. Similarly, it may be argued that we make ethanol from corn because the ethanol has an important use as motor fuel, and corn has been in surplus.20 Argued in a different way, the inputs of energy toward production of ethanol involve, for example, heat for distillation, which may be produced from coal-burning power plants or even by the burning of coal within the ethanol plant itself, and we need ethanol more than coal. However, as previously indicated, an inhibiting quality of coal burning is its implication for global warming.

Second, several studies have shown that use of ethanol as a motor fuel increases emissions of nitrous oxide precursors of ozone and air pollution, which are already serious causes of asthma and allergies in several U. S. cities. We note in passing that this matter is also controversial, but it appears that those who claim that ethanol reduces harmful emissions benefit personally from ethanol manufacture. Third, the fermentation process produces 2.7 gallons of ethanol per fifty-six pound bushel of corn. This means, for example, that two billion bushels of corn, about 20% of the U.S. corn crop, can produce 5.4 billion gallons of ethanol. Because the energy in ethanol on a volume basis is about 70% of that in gasoline, this is equivalent in gasoline to less than 4 billion gallons or 100 million barrels. As noted above, this is only five days of U.S. oil consumption! In these rough calculations, we see truth in part of a statement released by U.S. Senator John McCain (2003): ". . . ethanol does nothing to reduce fuel consumption, nothing to increase energy independence, and nothing to improve air quality". Regrettably, Senator McCain as a candidate in 2007 for the Republican presidential nomination in 2008 is now supporting the national ethanol program because the nature of the U.S. political system gives inappropriate power to interests that benefit from the program. Some other candidates for political office in the

United States have similarly switched their positions. A fourth aspect of the ethanol program is its impact on the availability of corn for feed, owing to diversion of a portion of the crop for manufacture of auto fuel. At an extreme, in reference to a perceived looming shortage of animal feeds and human food, the conversion of foods to fuels and especially the ethanol programs have been labeled "The Internationalization of Genocide" by the Cuban publication *Granma* (Castro, 2007). Strong general condemnation in this publication also notes the small fraction of fuel needs to be provided by conversion of large amounts of grain for "voracious automobiles". Certainly, the price of corn and other feeds is being increased by increased demand for corn and by planting to corn of land formerly used to grow other feeds. Between 1980 and 2006, the price of U.S. corn fluctuated considerably but, with few exceptions, remained below $2.50/bushel (56 pounds).

At this writing in September 2007, the price of corn is about $3.75 per bushel21 and this with other related price increases is receiving most of the blame for a reduction of U.S. food aid by more than half since the year 2000. This surge in the price of corn has a direct impact on the cost of animal feeds and hence on the price of beef, chicken, and pork, and newspaper articles have carried many indications of related concerns. This has carried over to demonstrations in Mexico, for example, since the price of corn relates directly to the price of tortillas, a dietary staple there that consists almost wholly of corn. However, in the United States, the impact on many items bought in stores may be minimal because the overwhelming part of the price of typical packaged foods reflects value-added processing and costs of packaging and distribution following purchase of the raw commodity. For example, a 14-ounce package of corn tortilla chips, which sold in U.S. supermarkets for about $2.35 in August 2007, contains less than 4.5 cents of farmers' share with corn prices at $3/bushel. A doubling of the price of corn would raise the price of the tortilla chips only 4.5 cents! Somewhat more significant would be the impact on a four-pound package of corn flour, selling for $2.50 in U.S. supermarkets. Farmers' share here is about eighteen cents, so a doubling of the corn price would raise the price to consumers by eighteen cents. Of course, these simple calculations do not

A possible positive international benefit of a higher corn price lies in improved competitiveness of corn grown by traditional methods in less industrialized countries. Thus, the historical low price of corn grown in the U.S. by industrial methods and exported to Mexico under the North American Free Trade Agreement has reduced the marketability of corn grown on small farms in Mexico, and this may change with a higher price of U.S. corn. Another small plus is distillers grain, the high protein product that remains after fermentation of starch. This product can be fed to cattle during the finishing stages of their fattening for slaughter in our industrial agriculture. A fifth important negative impact of both the ethanol program and biodiesel program (see below) is reduction of already stressed water supplies, especially in western United States. This concern has been widely publicized in the United States during fall 2007, and is treated in detail by the U.S. National Academy of Sciences (2007).

The U.S. ethanol program is subsidized at the federal level by a nominal tax credit of 51 cents per gallon, and further supported by a tariff on importation of ethanol. The tax credit has been shown in a report by the Congressional Research Service to amount in actuality to 68 cents/gallon owing to the manner in which the credit is administered (Congressional Research Service, 2005). The federal subsidy is augmented in Oklahoma by legislation granting an additional tax credit of twenty cents/gallon.

Often overlooked in the corn-to-ethanol program are heightened general negative impact of increased corn production on ecosystems and high cost of transporting ethanol. Land planted to corn in 2007 totaled about 93 million acres, the highest since 1933, and recent yield of about 155 bushels/acre is nearly double that typical of thirty years ago. The increased acreage is a response to the ethanol program and the increased yield reflects large fertilizer inputs, which involve energy-intensive production of fertilizer with dark implications for hydrocarbon inputs and emissions of greenhouse gases. There are also serious implications for erosion of land in increased production of corn, because soil erosion under corn far exceeds replacement.

Regarding transport of ethanol to markets, existing pipelines cannot be used because ethanol is a strong solvent and would become contaminated with pipeline residues while causing corrosion to the pipelines themselves. Therefore, pending solution to these problems, more expensive truck and rail transport is necessary, and these factors have not been accounted for in the federally supported program. The corn-to-ethanol program is also causing a large increase in the price of farmland, which, according to articles in The New York Times on August 10, 2007, is increasingly shutting out beginning farmers with limited capital. In spite of loan programs such as those provided by the U.S. Farm Service Agency, the average age of the U.S. population that actually farms has been increasing for years, and efforts to facilitate entry of young people to farming have been increasingly assumed by individual States and by such pro-bono organizations as the Center for Rural Affairs (2007).

Much touted is an ethanol program in Brazil, which provides about 25% of auto fuel there. During 2007, Brazil achieved independence from imported oil owing to a combination of its ethanol program with a significant discovery in an offshore oil field. Brazilian ethanol is made from sugar, which is easier to ferment than corn, and about 4 billion gallons is produced annually from sugar cane grown on about six million hectares of farmland (10% of farmland in Brazil). It is much easier to satisfy Brazil's automotive fuel demand than U.S. demand, because the area of Brazil is 8.5% larger than the U.S.' "lower 48" while its automotive fuel use is only 10% of that in the U.S. It has been reported that the farmland devoted to sugar cane in Brazil was formerly used to grow fruit and vegetables and that no appreciable amount of rainforest has been removed in order to accommodate demand for sugar (Lagercrantz, 2006). However, we wonder whether some of those displaced from horticulture will clear present jungle for new farms.

Finally, research toward conversion of cellulose to ethanol is now subject to much discussion. Will cellulosic conversion be a successor to corn-to-ethanol in the United States? Grasses, especially switchgrass,23 are commonly portrayed as a viable future rootstock, and development of effective conversion technology is widely publicized as imminent. Extensive research on this subject is underway, with the U.S. Dept. of Energy awarding hundreds of millions of dollars for development of pilot plant s for experimentation with several technologies. Knowledgeable botanists have expressed reservations, noting that serious implications of continuous monocropping for net energy consumption, pesticide usage, erosion, water use, and reconversion of the conservation reserve24 have not been well explored and that little note is being taken of the large amount of cellulose that would be required to replace just a few percent of current U.S. oil consumption. In short, some think that much use of switchgrass in a monoculture and other cellulose for ethanol production could produce an industry resembling that for corn, and some field experiments to clarify these issues are being planned (Wallace, 2007). Also to be considered is the impact of a cellulosic industry on maintenance of domestic herbivores.

BIODIESEL

In a diesel engine, the fuel air mixture is compressed so much that the accompanying rise of temperature causes self-ignition of the fuel. The higher compression and temperature in a diesel engine than in the usual internal combustion engine produces higher fuel efficiency, i.e., increased mileage with a vehicle. Diesel engines are desirable for this reason, and also because of their simplicity associated with absence of spark plugs and distributor. Diesel fuel is less volatile than gasoline and can be made from both petroleum, with declining availability, and from animal and plant fats and oils. Diesel fuel with a recent biological origin is known as biodiesel. Glycerin in animal and plant fats and oils must be removed before the lipids can be used in diesel engines. The usual refining process, known to chemists as transesterification for removal of glycerin, involves a reaction of the oils with an alcohol, addition of some water and later heat to remove the water, and various other stages including addition of catalysts, which are recovered for reuse. Several somewhat similar refining processes can be used effectively and are well established. The removed glycerin has a market in soap manufacturing and a few other applications.

In the United States, soybeans are presently the principal source of the oils used to produce biodiesel, and about 73 millions acres of cropland have been devoted to soybean production. With a generous estimate of production at 40 bushels (each 60 pounds)/ acre, and 1.4 gallons biodiesel/bushel, total biodiesel production would be 4 billion gallons or 100 million barrels, if all of the soy beans now grown in the U.S. were used for oil production. Corresponding to the preceding analysis of ethanol, this would replace only the amount of petroleum that the U.S. presently consumes in five days! Note that strongly negative conclusions are implied even though no consideration has been given here to other negatives associated with energy inputs required to grow, harvest, and transport soybeans.

In parallel with the ethanol analysis, large-scale production of diesel fuel from oil seeds would have unintended undesirable consequences on markets and the conservation preserve. It is reported by George Monbiot25 that a rush to produce biodiesel from palm oil in Malaysia is causing great losses to primitive rainforest that already represents only a remnant wildlife habitat. The palm oil industry is cognizant of such criticisms and is planning a conference during fall 2007 to review its practices.

Other sources of biodiesel are waste oil at restaurants and homes and animal fat produced by meat processors. During summer, 2007, Tyson Foods, Inc. announced contracts that would produce 175 million gallons (four million barrels) of biodiesel fuel from 25% of Tyson's fat production. This is just 20% of U.S. petroleum supply for one day. While having little effect on U.S. petroleum dependence, this diversion is expected to cause significant price rises in the soap industry. In the presence of serious unintended consequences and far-flung ripple effects, production of biodiesel fuel in the U.S. is presently subsidized by $1 per gallon at the federal level and also receives subsidies in many States, including 20 cents/gallon in Oklahoma.

Actual U.S. production of biodiesel from soybeans in 2007 is about 300 million gallons or seven million barrels per year, about a third of petroleum products used in one day. In April 2006, a biodiesel facility was opened in Durant, Oklahoma, and was slated to sell the soybean derivative under the brand name, BioWillie, given by famous singerWillie Nelson. On July 13th, 2007, it was reported that a group of note holders had filed an involuntary

Chapter 7 bankruptcy petition against the Dallasbased owner of the biodiesel production plant. According to the petition filed in the U.S. Bankruptcy Court for the District of Delaware, Earth Biofuels had not been paying its debts as they became due. However, on January 21, 2008, it was reported that the petition for involuntary bankruptcy had been dismissed by the Court and that Earth Biofuels had consummated an agreement with Alliance Processors to purchase waste grease collected at restaurants in Texas. Up to 400 thousand gallons of grease per month is expected to be supplied.

This is a commendable program. After all, "Waste not, want not", but it should be recognized that if each gallon of grease makes nearly a gallon of diesel fuel, the grease collection is equivalent to about nine thousand barrels per month, or less than one-twentieth of one percent of the petroleum used in the United States each day! The first article quoted the Chair and CEO of Earth Biofuels, "The biofuels industry and other alternative fuels are absolutely essential to our nation's energy security and our ability to maintain economic independence. The goal of energy independence won't be achieved through use of a single technology." Your present author does not agree with the first part of the quote but believes that the second part is probably true.

Finally, algae are still another source of biodiesel, theoretically very promising. Again, however, the practical challenges are very great and the ultimate outcome of research in this area is speculative. At this writing, several companies are involved, and Greenfuel Technologies of Cambridge, Massachusetts, U.S.A., is partnering with Arizona Public Service at one of the latter's power plants to develop a system that would feed algae with the plant's carbon dioxide emissions.

GREENHOUSE WARMING AND ITS CONNECTIONS

We discuss global warming because it carries grave implications for the future human condition and because it is being caused by human activities, mostly by the burning of carbon-containing fossil fuels for transportation and for generation of electrical power.26 The global warming issue is thus tightly connected to the fueldecline issue as illustrated in a short article by your author (Kessler, 1991) and elsewhere. Extraction and burning of carbon fuels since the start of the industrial revolution and particularly the burning of a substantial fraction of extractible resources since World War II has been the source of the present developed economies with high levels of material well-being, and naturally there is a wish to preserve and enhance this condition. In this connection it is important to have in mind both the relative amounts of carbon dioxide produced by the combustion of basic fuels and their heats of combustion. With each million BTU produced by combustion of carbon, about 119 kilograms of carbon dioxide are produced. Coal used as fuel is the largest emitter of carbon dioxide in relation to energy produced. Continued present political reality carries implications for changed weather and climate, for rapid changes in agricultural practice, for substantial rises of sea level, and for changes of oceanic flora and fauna in response to oceanic uptakes of carbon dioxide with resultant increase in oceanic acidity.

There are at least four aspects of global warming with public interest. First, How enduring will global warming be as presently measured? Second, Are human beings a principal cause? Third, Is global warming important for human beings? Fourth, Can it be mitigated by humankind? Although global warming continues to have outspoken deniers in 2007, a proper answer to each of these questions is a resounding "yes", but it is necessary to add that mitigation of global warming and its effects presents to humankind a challenge unprecedented in its magnitude. It is not at all clear at this writing that the challenge will be well met.

THE REALITY OF GLOBAL WARMING

The Intergovernmental Panel on Climate Change has issued impressive documentation (more than fifteen hundred pages) including both technical details and accounts readily understood by laypersons. These accounts are available on the internet (IPCC, 2007) and are an excellent source of details concerning the following account.

Global warming is world-wide and given the immense variability of weather, no local phenomenon taken by itself proves or disproves global warming. Aggregation of many local effects can be evidence of global warming. The temperature record at Oklahoma City from January 2004 is a small piece of evidence for global warming. Thirty-four of forty-eight months from January 2004 through December 2007 had above-normal temperatures and thirteen experienced below normal. The largest above normal was +11.0Fahrenheit and the largest below normal was –4.6F. The overall average was +1.93F above normal. Also, during this period, 29 high temperature date records were tied or broken (either the maximum temperature for a particular date or the highest minimum temperature for a particular date) and five low temperature records were tied or broken. Fortunately for the local inhabitants, while Oklahoma winters during 2004–2007 tended to be mild, summers there, usually very hot, were cooler than the long-term average in 2004 and 2005 and not excessively warmer than average in 2006 and 2007.

Sometimes skepticism about global warming is produced by other extreme local conditions. Such was especially the case during the weekend of April 7–8, 2007, in North America, when a severe cold wave covered eastern sections and some low temperature records were broken. However, a figure from the National Oceanic and Atmospheric Administration that depicts temperature anomalies over the whole of

Earth (NOAA, 2007) shows that our planet as a whole was experiencing above normal temperatures at that time. With a few minor exceptions, eastern North America was the only place on this third planet from Sun that was experiencing temperatures substantially below seasonal averages, and temperatures well above long-term averages prevailed over most of Earth, especially in Arctic regions. And almost all global anomaly charts during 2006 and 2007 are similar in showing a larger area of Earth with above normal temperatures than with below normal temperatures. Of course, pattern details change constantly.

Another indicator of global warming is in a report from the National Oceanic and Atmospheric Administration on March 15, 2007. This states that overall on Planet Earth, the average temperature during three winter months in the northern hemisphere, December 2006 through February 2007, was the warmest recorded since such record keeping began a little more than one hundred years ago. And the eleven warmest years of record on a global basis have occurred during the past twelve years.

Consider conditions in Europe. In an article (*Weather*, 2007) published by the Royal Meteorological Society in the United Kingdom, it is stated that the 12-month period from March 2006 through February 2007 was the warmest ever recorded in the 350-year period of the central England temperature (CET) record. The CET record is the longest instrumental temperature series on Planet Earth. Furthermore, records during the past several years, documented monthly in *Weather* show that practically every month has had above normal temperatures overall in both the U.K. and in continental Europe, and readers will well remember the heat wave of summer 2003 in Europe, when up to 35 thousand deaths were attributed to record-breaking high temperatures. There were heat waves in Europe in 2006 and 2007 also, though of lesser intensity (but 45C in Greece and some Balkan states, with devastating forest fires in 2007), and there has been a substantial increase in the frequency of heat waves in Europe.

A report in EOS (Komar, 2007) documents a convincing increase of wave height since 1985, as measured by buoys near the southeastern coast of the United States. The increased wave height is presented as indicative of increasing storm intensities, a consequence of rising ocean temperatures. There has also been technical documentation indicating increased frequency of drought and flood, and possible increased frequency and severity of hurricanes. Flooding in central England during summer 2007 and record-breaking floods in parts of India during 2006 and 2007 are not proof of global warming, but are suggestive. During August 2007, observations showed that Arctic sea ice had retreated to a record minimum. Melting was particularly prominent north of the Arctic coasts of Alaska and Siberia. By September 2007, the Arctic ice limit had retreated northward at some longitudes more than 500 miles further from its distance from the Siberian coast on same dates in 2006,29 much more than expected. In this connection, a chapter by Hansen et al., (2007), seems important. For about 20 years, Hansen has been a principal spokesperson for the climate change science community. In its indications that the

IPCC documents are conservative estimates of the rate at which climate change is proceeding and of the rate at which remedial action must be taken to avoid passage of a point of no return, this chapter presaged the remarkable 2007 retreat of Arctic ice.

CLIMATIC FLUCTUATIONS

There are several causes of climatic fluctuations. Diminution of solar radiation during the Middle Ages is thought to have contributed to global cooling at that time, the so-called Maunder Minimum. Earth's orbit and inclination to the ecliptic are perturbed by the gravitational influence of other planets, particularly Jupiter and Saturn, as analyzed by Milankovich 100 years ago, and some of the major historical ice ages and subsequent warmings are attributed to these variations. Volcanism with strong emissions of carbon dioxide and particulates are believed to influence climate, with particulates tending to reduce temperature and carbon dioxide tending to increase it. Depending on the cause, some climatic fluctuations are opposite on northern and southern hemispheres, and some are synchronous.

Concerning the present climatic fluctuation, it has been shown that geothermal heat associated with volcanic eruptions, black smokers on the ocean floor, etc., are not contributors (Roach, 1998). And although some blame solar variation for climate change, the present oscillation with the sunspot cycle is less than 2 watts/meter2 in a total radiance of 1370 watts/meter2 and cannot be a significant factor. Present concerns are principally related to carbon dioxide, which, next to Sun, of course, and water vapor, is the principal regulator of temperature on Earth.30 The heat trapping effects of carbon dioxide have been known for at least one hundred years, and were well taught at MIT and elsewhere fifty years ago. Increase of atmospheric carbon dioxide causes a diminution of heat transfer by radiation from the lower atmosphere to the upper atmosphere, an increase of temperature in lower atmospheric layers, and a compensating increase of heat transfer by atmospheric convection (mass motion, as in water boiling on a stove).

Climatic temperature fluctuations during the past 850 thousand years have been deduced from analysis of ice cores obtained in Greenland and Antarctica. Atmospheric gases in the ice essential to these analyses include carbon dioxide and oxygen isotopes 18O and 16O. Particulates are also in the ice, which shows annual striations. It is important that although the maximum atmospheric carbon dioxide content during this historical period was about 300 parts per million by volume (ppmv), the 2007 content is about 385 ppmv and increasing by about 2 ppmv each year.

Precision measurement of atmospheric CO2 was begun by Charles Keeling in 1958, and is now monitored at stations around the world. Records show annual increase every year and a within-year variation that is attributed to the cycle of plant growth in the northern hemisphere. During the 1960s, annual increase was only about one half ppmv per year, and the four-fold rate of increase since then corresponds closely with the increasing rate at which humans are burning fossil fuels. Globally, carbon burning has increased from about two billion tons annually during the late 1950s to more than seven billion tons annually today, with total present-day emissions of carbon dioxide about 25 billion tons annually (Marland, et al, 2005). U.S. facilities that generate electrical power typically burn every day all the coal contained in railroad trains of even more than one hundred cars. Each car may contain about seventy tons of carbon in coal, and each power plant thereby produces about thirty thousand tons of carbon dioxide every day.More coal-burning power plants are being built here and elsewhere, one per week in China, where there are awesome environmental consequences of its rapid industrialization (Kahn and Yardley, 2007).

As noted above, the 2007 carbon dioxide content of Earth's atmosphere is about 385 ppmv, 30% above 300 ppmv, which was the approximate maximum during the pre-industrial 850 thousand years for which atmospheric values can be accurately determined by analyses of gases trapped in polar ice. The present extraordinary content of carbon dioxide is believed to be the significant cause of rising global temperatures.

POLITICAL AND SOCIAL CONDITIONS, ESPECIALLY IN THE UNITED STATES

Political and Social Conditions in the United States are determinants of all of the legislation passed in the U.S. Congress and in state Legislatures. Of course, we are here concerned with legislation related to U.S. dependence on foreign suppliers for energy and the intertwined problems of global warming and agriculture. Very regrettably, serious deficiencies in rational attention to science, to unintended consequences, and to long-term issues are prominent in the politics of the U.S. government. The shape of legislation is very much determined by moneyed interests that work through lobbyists. Lobbying is an important and needed source of information, but it seems beyond proper control in the United States. Numerous publications from pro bono organizations such as the October 2007 issue of *National Voter* from the U.S. League ofWomen Voters inform the public of moneyed and corrupt influences that hurt this country, but public power and even public interest are so far inadequate to stem related bad practice sufficiently. Much of the U.S. public seems focused on entertainment. Even the U.S. President, though faced with a war in Iraq, said, "Go to Disneyland". A scholarly and comprehensive discussion of the U.S. political system (and some other systems) has been presented by Vago (1981).

As previously noted, the United States is the world's largest emitter of carbon dioxide, just a bit ahead of China, which is nearly caught up in the year 2007. Although large emitters, both the U.S. and China have been among the least inclined to control their warming emissions. China notes that although it will soon pass the United States in total emissions, its per capita emissions are only about one-third those of the United States, and the average standard of living of its people lags seriously. Compounding the condition of present large emissions, there are continuing strident calls in the United States and elsewhere for further economic growth, which can only increase demand for electrical energy and liquid fuels, both of which associate with increased emissions of carbon dioxide. If physical growth and associated demand growth continue, ultimate demand for fuels would increase along with the emissions therefrom, regardless of any measures directed toward conservation or improved efficiency.

In the United States, search for replacement of petroleum-derived liquid fuels reflects ardent wishes to preserve and even continue to enhance the automotive economy. The search involves investigation of alternative fuels as described in previous sections of this chapter and recovery of energy resources via activities made economically feasible by the high and rising price of traditional sources. Thus, for example, there are immensely expensive oil recovery projects in deep waters of the Gulf of Mexico, where oil rig leases now cost up to a million dollars daily. Extraction of oil from tar sands in western Canada is of special concern. This industry produces substantially more carbon dioxide per unit of oil that is extracted, refined from its tarry beginnings, and delivered to users than the traditional oil industry. As traditional liquid fuels become scarcer, there are also calls for their production from coal and natural gas. This would also enhance emissions of carbon dioxide.

Sequestration (permanent burial) of carbon dioxide for sufficient reduction of global warming is costly. There has been considerable discussion of sequestration in the U.S. press, but the only significant practice, so far, occurs where injection of carbon dioxide enhances recovery of petroleum (tertiary recovery). Proposals to reduce carbon emissions through a tax on carbon burning have been implemented in a few European countries and others, but not in the United States, owing to opposition from special interests. Similarly, although the cost of limiting mercury emissions from coal has been reported to be less than 0.3 cents per kilowatt-hour, installation of such emission control is being implemented on a time line longer than ten years (Srivastava, et al. 2006).

Resistance to change is illustrated in the U.S. State of Oklahoma with striking examples of efforts to continue to expand highway travel while ignoring opportunities for provision of improved public transportation and freight service via rail, which is much more energy-efficient and if better utilized would significantly reduce both the threats of global warming and the dependence of the United States on petroleum. For example, powerful highway interests have been intent on replacing the Oklahoma City Crossstown Highway (U.S. Interstate Route 40) at a cost of more than half a billion dollars for less than four miles of new road. This program, started in the mid-1990s, proposes a new large highway that is not a public need on a route that would destroy the Union Station rail yard, owned by Oklahoma City. Union Station, in excellent condition and on the U.S. Historical Register, was a multimodal transportation center fifty years ago, and was purchased in 1989 for announced use for public transportation. Tracks and rights of way to all parts of the State are owned by the State and converge at Union Station, although all tracks are not in good condition at this writing.

The Crosstown replacement proposal illustrates the immense power of the U.S. automobile and truck lobbies and related special interests. If implemented, this proposal would increase truck travel through Oklahoma City and increase ozone and related health problems there while reducing prospects for economical, energy efficient public transportation and freight service throughout the State. This would occur with Oklahoma already behind many other U.S. states and cities in provision of public transportation. While many Oklahomans can hardly afford to buy and maintain the private cars necessary there for travel to work or cannot drive for reasons of health, a variety of other reasons encourage efficient transportation of passengers and freight by rail, and retention of a facility that could be a hub for both freight and passenger service as it once was.

Lack of sufficiently effective programs in the United States is also a consequence of a cultural condition described in the *Harvard Divinity Bulletin* (Weiskel, 1990). Many individuals think in terms of anthropocentrism, e.g., "Earth is made for Man", and policies along this line are too often manifested in government. We should also be concerned with exceptionalism, the notion that humankind, owing to large brains, is exempt from the laws of nature applicable to other living beings. Both concepts have a basis in the Abrahamic religions established in both western and Islamic societies. Culture wars in the U.S. and elsewhere often pit these traditional concepts against new ideas about humankind's proper place. The new ideas spring partly from a torrent of new science about the cosmos ranging from the infinitesimal to the farthest galaxies. Regrettably, the new ideas also bring a kind of new religion with a new exceptionalism. The new religious beliefs hold that problems as they develop will inevitably be solved by new science and technology, and some government support of research stems from this attitude. Indeed, in speeches from natural limitations. Much legislation provided by the political system in the United States is an exchange for financial contributions to campaigns. Will our system (and others, too) remain inadequate to deal with the global warming and energy decline phenomena? If it does remain inadequate, it will not be because the U.S. system is vastly different than it used to be – although there has been concentration of control of media by narrow interests, this control over news delivery has been somewhat offset by democratizing effects of the Internet. Historically, our political system has frequently supported powerful groups that sacrifice the good of a large sector for personal short-term benefits. This author thinks that the last times that populace anu government rose to needed heights was when the critical nature of conditions related toWWII became more than obvious. And subsequent toWWII there was the good Marshall Plan.

In the United States and elsewhere, many research programs are well funded. As noted in a short article (Kessler, 1991), the political establishment is pleased to provide the wherewithal, in part because the hope for favorable outcomes is a basis for postponement of actions that are politically difficult to implement even though they could be immediately effective. And, of course, research must be encouraged; a plethora of research outcomes in every field of study are the principal basis for our industrial and postindustrial worlds, and further highly favorable results seem inevitable.

For example, a recent helpful outcome in Japan has produced light emitting diodes (LEDs) that are about 50% efficient in their production of light from electrical energy, and the cost of LED production is being reduced rapidly. LEDs may be on track to replace both incandescent lights with efficiency about 5% and fluorescents, 25%. The U.S. Dept. of Energy has estimated that about 22% of electricity production is devoted to lighting, so the new products may lead to both reduced CO2 emissions and better lighting around the world, including in communities remote from utility power (Ouellette, 2007).

Important developed differences between now and decades ago are more in the nature of our times than in qualities of our political system. General demand has risen and continues to rise with increasing world population, and some basic resources that are essential to maintenance of infrastructure and provision of essentials are not as plentiful as formerly and are more expensive to obtain. The immense power of tools created by spectacular advances in science and technology means that malfeasance in the application of those tools leads to increasingly harmful consequences. Thus, private automobiles have provided unprecedented and very welcome mobility to many, but they are still being promoted even though they are principal contributors to carbon dioxide emissions and decline of liquid fuels. While products of advanced science and resultant technologies are essential to most of our daily lives, many more people in the United States than in Europe seem to reject findings and implications of science when those findings conflict with historical matters of faith or call for specific short-term sacrifice for dimly-perceived benefits in the long term.

Science and technology are seen as the major source of means for tapping the wealth of Earth. To what extent may further advances lead to means for marked reduction of our impacts? Such favorable developments will depend much more on scientific guidance to research directions than on political guidance! Geometrical orientations of Earth to Sun are projected to rule out global cooling and recurrence of glaciations for another 30 thousand years, and this means that global warming will continue inexorably unless emissions of greenhouse gases are greatly diminished or there is an unexpected diminution of Solar radiation or extensive volcanism on Earth. Therefore, it may well be that within a few decades, humans on Earth will have to accommodate powerful forces that will make early adjustments seem easy by comparison. New problems may well include migrations of millions of people forced to leave submerging habitats, shortages of water in areas now dependent on glacial runoff, hotter summers, fluctuations of food supply following intensified droughts and floods, and increased social unrest. There are solutions to global warming problems, but none is easy, and most political systems are inhibiting.Will we humans meet this immense challenge to our established ways and cultures? Delay compounds difficulty and cost of solutions.

CONCLUSIONS

The United States has not yet a single program effective toward reduction of its dependence on foreign sources for liquid fuels or toward mitigation of the looming disaster represented by global warming. If existing programs were effective, we would expect that imports of petroleum products would be declining, but such imports are continuing to increase. And the existing biofuels programs are already damaging the agricultural economy. In large part, the programs in place are a consequence of a political system whose legislation is too-much based on contributions from the already rich and powerful, and insufficiently responsive to conditions and findings from advanced and still burgeoning science and technology. Overall, the situation is a consequence of the human condition, little changed during thousands of years.

Such programs as improved insulation of existing houses, new construction of "green" buildings, and facilitation of transportation alternatives such as bicycling, are steps in right directions and have won grass-roots support, but all are far too small. The major programs, ethanol from corn and sugar cane and biodiesel from palm oil, soybeans, and canola are deceptive responses. They provide short-term profit to special interests and they do provide fuels, but even the aggregate amount of fuels produced in these programs is a trivial proportion of present consumption and, the production processes yield, at best, no net reduction of carbon dioxide emissions. The alternative fuels programs damage the agricultural economy by causing increases in the price of corn and other human foods and livestock feeds, losses of already diminished habitat including tropical rainforests and wildlife, and losses of topsoil and increased stress on water supplies. As noted above, unless carbon dioxide emissions are quickly reduced, global warming will be a very serious matter for future generations and will force large adjustments in ecosystems worldwide. Concern rises because in the United States and in rapidly developing countries such as China and India, policies remain strongly oriented toward economic and even physical growth with increasing emissions of carbon dioxide.

What should be done in the United States, for example, beyond such programs as tightening CAF´E35 standards, weatherizing homes and utilizing energy-saving construction in new work, installing solar heating, and expanding use of time-of-day pricing of electricity, all of which are or would be good though inadequate? A proper practical course is difficult to identify, and an effective course may be impossible to identify. In other words, it may be too late to avoid serious damages from global warming and to preserve social order in face of fuel declines. But, we must keep trying, and it is clear enough that in order to confront consequences of global warming and decline of liquid fuels, societies in developed (and developing) countries must practically be turned on their heads! And if they do not turn themselves soon, they will be turned later by large forces beyond human control.

As a first step, the notion of continuous economic growth must be abandoned,36 and global population, which has increased threefold in your author's lifetime, must be much reduced. Whatever else is done, if population growth proceeds, all other saving actions will be nullified and even overwhelmed owing to increased demand. Abplanap's succinct statement (1999) applies, necessary changes being made, to physical growth of many entities in the presence (or absence) of technological advances: ". . . Any kind of agricultural 'green revolution' which is not accompanied by effective population control merely resets the limiting parameters at higher levels and enables countries with a large proportion of starving citizens to increase the absolute numbers of starving people".

Is population reduction feasible? Population is sustained with an average birth number near 2.1 per female inhabitant. If this average were reduced to 2.0 the impact on individuals would be very minor but the eventual impact on world population would be major. If world population were to decline just one percent per year, numbers would be reduced by half in 70 years and again by half in another 70. In 2007, this must be seen as only a utopian dream, since the large proportion of young people in the present world population guarantees substantial growth of the global population in the near term. Further, strong diverse forces, even the U.S. government at this writing, offer little or no support for birth control,38 and Chambers of Commerce all across America promote growth among the highest of their priorities. Of course, population matters are very different in different economies, demographies, and cultures, and associated problems, including treatment and education of females, are not explored here.

Second, it would be helpful in the United States to have a massive shift in funding from highway building to construction of a national rail system for both passenger travel and improved freight transport. Such a system, emulating that already in place and still under rapid development in Europe and somewhat too in Asia, would be inherently more energy efficient than automobiles and truck travel on highways, and even further emission reductions would be achieved to the extent that trains become more fueled with electricity from overhead wires or from liquefied natural gas in place of diesel fuel.

Such a transportation alternative in the U.S. might be paid for in part by an increased federal tax on gasoline and diesel fuels. If rail were more emphasized, U.S. highways would be less burdened with cars and trucks, highway maintenance costs would decline, and emissions of carbon dioxide and health-threatening gases from the automotive sector in this leader country would decline. And decline of truck traffic would quicken if trucks were taxed in relation to the maintenance costs they impose – road damage is proportional to the fifth power of axle weight.40 Groups of citizen-activists are working in these directions, but during 2007 in the United States, there is little official interest in such programs – indeed, such programs lack substantial support from the federal level in the United States and are opposed by highway and automotive lobbies. In 2007 there is still strong political support toward expansion of the highway system.

Third, further enhancement of already burgeoning communication technologies may proceed to a level that somewhat reduces energy-consumptive travel. The three items above could be resource-conserving approaches in a relatively short term. But for true sustainability in terms of geological age, we should, barring success with nuclear fusion as a source of electrical energy, begin to explore development of a very broad solar economy, because only solar energy is projected to endure much as at present for billions of years. This means that solar power plants would be built with help from fossil or nuclear fuels to support an economy with fewer human numbers indefinitely, and the solar power would be used to maintain and enhance the power system itself. This vision of a farther future is mentioned by Patzek on his website and a possible solar path has been detailed by Zweibel, et al. (2008).

So, in summary, What is our food and fuel future? It is highly problematic, and a decent future for humans is much dependent on rationalization of decision-making at all levels to findings and implications of science and technology! The rapid pace of change in this 21st century also calls for a much more rapid response of proper decision making to major findings of science and technology.

Will humanity on Earth be a "flash in the pan"? Consider a 30-volume encyclopedia, each volume with one thousand pages, each page with an average one thousand words. Let these thirty volumes present a linear history of Life on Earth since multi-celled organisms became prevalent perhaps one billion years ago, with the start of accumulation of the fossil fuels that we humans use today. How much space is devoted to the sixty-five years sinceWorldWar II, during which we humans have extracted about half of Earth's readily extractable liquid fossil fuels and much coal, and caused an astonishing increase in atmospheric content of carbon dioxide? Is the answer disturbing? Only two words on the last page of the last volume! How long will we endure and how much space might describe our future post-industrial society?

9

A Framework for Energy Alternatives

Standard economic analysis does not accurately account for the physical depletion of a resource due to its reliance on fiat currency as a metric. Net energy analysis, particularly Energy Return on Energy Investment, can measure the biophysical properties of a resources progression over time. There has been sporadic and disparate use of net energy statistics over the past several decades. Some analyses are inclusive in treatment of inputs and outputs while others are very narrow, leading to difficulty of accurate comparisons in policy discussions. This chapter attempts to place these analyses in a common framework that includes both energy and non-energy inputs, environmental externalities, and non-energy co-products. We also assess how Liebig's Law of the minimum may require energy analysts to utilize multi-criteria analysis techniques when energy may not be the sole limiting variable.

INTRODUCTION

Human energy use, ostensibly the most important driver underpinning modern society, may soon undergo a major transition of both kind and scale. Though numerous energy technologies are touted as alternative supplies to fossil fuels, scientists and policymakers continue to lack a meaningful and systematic framework able to holistically compare disparate energy harvesting technologies. Net energy analysis attempts to base decisions largely on physical principles, thus looking a step ahead of political and/

or market based signals distorted by fiat monetary data. The importance of net energy has been overlooked, primarily as a result of confusing and conflicting results in energy literature. In this chapter, we (a) provide an introduction to the history, scale and scope of human energy use (b) reiterate the role of net energy analysis in a world of finite resources, (c) establish a two dimensional net energy framework synthesizing existing literature and (d) illustrate (via the example of corn ethanol) why multi-criteria analysis is important when energy is not the only limiting variable.

NET ENERGY ANALYSIS

Energy, along with water and air, completes the trifecta of life's most basic needs. Organisms on the planet have a long history of successfully obtaining and using energy, mostly represented as food. Indeed, some have suggested that the harness of maximum power by both organisms and ecosystems from their environments is so ubiquitous it should be considered the Fourth Law of Thermodynamics (Odum 1995). Cheetahs, to use one example, that repeatedly expend more energy chasing a gazelle than they receive from eating it will not incrementally survive to produce offspring. Each iteration of their hunting is a behavior optimized to gain the most energy (calories in) for the least physical effort (calories out), thus freeing up more energy for growth, maintenance, mating and raising offspring. Over evolutionary time, natural selection has optimized the most efficient methods for energy capture, transformation, and consumption. (Lotka 1922) This concept in optimal foraging analysis extrapolates to the human sphere via *net energy analysis,* which seeks to compare the amount of energy delivered to society by a technology to the total energy required to transform that energy to a socially useful form. Biophysical minded analysts prefer net energy analysis to standard economic analysis when assessing energy options because it incorporates a progression of the physical scarcity of an energy resource, and therefore is more immune to the signals given by market imperfections. Most importantly, because goods and services are produced from the conversion of energy into utility, surplus net energy is a measure of the potential to perform useful work for social/economic systems.

AN INTRODUCTION TO EROI – ENERGY RETURN ON INVESTMENT

Knowing the importance of energy in our lives, how do we compare different energy options? Unfortunately, the word 'renewable' does not automatically connote 'equality' or 'viability' when considering alternatives to fossil fuels. In assessing possible replacements for fossil fuels, each alternative presents special trade-offs between energy quantity, energy quality, and other inputs and impacts such as land, water, labor, and environmental health (Pimentel et al. 2002, Hill et al. 2006).When faced with these choices, energy policymakers in business and government will require a comprehensive and consistent framework for accurately comparing all aspects of an alternative fuel. Many criteria have historically been used to assess energy production technologies based on both absolute and relative yields and various costs (Hanegraaf et al. 1998). Many assess economic flows (e.g. Bender 1999, Kaylen 2005) while others focus on energy (e.g. Ulgiati 2001, Kallivroussis et al. 2002, Cleveland 2005, Farrell et al. 2006) or emissions (e.g. EPA 2002). With the recent acceptance of global climate change as a problem, energy analyses favoring low greenhouse gas emissions are becoming more frequent (Kim and Dale 2005, Chui et al. 2006). Though not yet widely accepted by market metrics, some other analyses have attempted to include environmental and social inputs as well as energy costs. (e.g. Giampietro et al. 1997, Hanegraaf et al. 1998, Pimental and Patzek 2005, Reijnders 2006).

The objective of an energy technology is to procure energy. A common measure combining the strength/quality of the resource with its procurement costs is the ratio of energy produced to energy consumed for a specific technology/source. This concept has many labels in energy literature including the energy profit ratio (Hall et al. 1986), net energy (Odum 1973), energy gain (Tainter 2003), and energy payback (Keoleian 1998). In this chapter, we focus on Energy Return on Investment (EROI) (Hall et al. 1986, Cleveland 1992, Gingerich and Hendrickson 1993) EROI is a ratio and is equal to 'net energy +1'. Total energy surplus is EROI times the size of the energy investment, minus the investment. We will use the terms energy gain, net energy and EROI interchangeably, throughout this chapter.

Humans and Energy Gain

Ancestral humans first major energy transformation came from the harnessing of fire, which provided significant changes to daily tribal life by providing light, warmth and eventually the ability to work metals, bake ceramics, and produce tools. (Cleveland 2007). More recently, the energy gain of agriculture further transformed human culture. Though the per unit energy gain of widespread agriculture was actually lower than many hunting and gathering practices, a large amount of previously unused land was brought under cultivation, thus freeing up substantially larger energy surplus for society as a whole. (Smil 1991) This is a first example of how an energy return combines with scale to determine an overall energy gain for society. Much more recently, the development of the steam engine catapulted mankind into the fossil fuel era by leveraging the embodied energy in coal deposits. The high energy gain of coal rippled its way through the economy akin to a deposit in a fractional banking system, and the industrial revolution had its first power source. In the 19th century, modern humans learned to unlock the hydrocarbon bonds in the higher quality fossil fuels of crude oil and natural gas, freeing up orders of magnitude more energy than our evolutionary forbears even dreamed about. The changing size of this subsidy, how to measure it and meaningfully compare it to potential energy substitutes that will be required to power future society is the subject of this chapter.

Current Energy Gain

The current scale of our energy gain is unprecedented. When coal, oil and natural gas are included, the average American uses 57 barrel of oil equivalents per year (BP 2005). Each barrel of oil contains 6.1178632×10^9 Joules of energy. An average man would need to work about 2.5 years to generate this amount of heat work1. Multiply it by 57, and the average American uses a fossil fuel subsidy equal to over 150 annual energy slaves. But the quality of oil is also fantastic – liquid at room temperature and highly dense – oil possesses energy quality that human labor cannot. An important nuance underlying the concept of net energy analysis, is that fossil fuel production is itself cannibalistic, as oil production uses a great deal of natural gas (and some oil) to procure. Coal production, wind turbine creation, solar photovoltaic panels, etc. all require liquid transportation fuels to generate their products in a modern economy. In fact, over 90% of world transportation is accomplished using liquid fuels. (Skrebowski 2006).

The scale of remaining recoverable crude oil is a topic under much debate, with many analysts saying we are already past peak production (Deffeyes), and others (IEA, Cambridge Energy Research Associates) saying we will reach a broad plateau by 2030–2040. A large number of analysts believe a peak in oil production will occur sometime in the next decade. However, few if any of these analysts look at how much of future oil and gas production nets down to the societal use phase after the energy costs have been accounted for. Nor is there a distinction made in 'crude oil' statistics between actual crude oil, ethanol, coal-to-liquids, etc. all of which not only have disparate energy costs, but different BTU contents as well.

The Hubbert curve of resource extraction is roughly Gaussian in shape, and the energy surplus (or lack thereof) drops down dramatically after its peak (see Hall et al., 1986 for an example on Louisiana). If oil is peaking soon, asking how much is still in the ground is not the most important question. How much can be brought to market at one time? How much energy is left after energy companies use what they require internally to procure the harder to find, deeper, more sulfurous, more environmentally and socially sensitive drilling locations, etc.? These questions ultimately address how much of our remaining fossil resources will be available for non-energy, non-government society.

An Energy Theory of Value

There is a rich history over many decades of the concept of an energy theory of value, dating back to Howard Scott and the Technocrats who stated that 'A dollar may be worth – in buying power – so much today and more or less tomorrow, but a unit of heat is the same in 1900, 1929,1933 or 2000' (Berndt 1983). In the 1970s, Senator Mark Hatfield argued that 'Energy is the currency around which we should be basing our economic forecasts, not money supply.' His efforts resulted in the passing of (now defunct) Public Law 93.577 which stipulated that all prospective energy supply technologies considered for commercial application must be assessed and evaluated in terms of their 'potential for production of net energy'. (Spreng 1988) And in a still broader sense, ecological analysts have long stated that money does not properly account for externalities – ecologist Howard Odum stated 'Money is inadequate as a measure of value, since much of the valuable work upon which the biosphere depends is done by ecological systems, atmospheric systems, and geologic systems.'

Why is Net Energy Important?

This 'work' Professor Odum alluded to requires an energy surplus. (Odum 1994) In a world where energy is likely to become scarcer, net energy analysis is more forward looking than conventional economic analysis, and as such can be an important tool for policymakers. Net energy is important because we need energy to accomplish work. The surplus energy of a system, or society, is what allows it to continue growth, maintenance, repair and leisure. Energy technologies can be stock or flow based. Stocks are depletable and non-renewable on human time scales. Flow-based resources are renewable, provided the infrastructure that supports them is renewable. There is only so much low entropy energy present in fossil fuel stocks and solar/tidal flows that can be accessed at a meaningfully positive energy return. If society has collectively become dependent on a certain aggregate energy gain system and attempts to replace it with a lower energy gain portfolio, while keeping all other inputs equal, then a larger % of societies resources (labor, capital, land, water, etc) would have to be devoted to energy procurement, leaving less available for hospitals, infrastructure, science, etc.

So in one sense, the Energy Return on Investment is a story of demand, and how a civilization uses their BTU endowments. A doubling in efficiency of use, or a doubling of conservation efforts, are equivalent to a doubling of an energy surplus. But if efficiency and conservation do not occur, we are left trying to maintain a *high gain system from new energy supply as original stocks of resources deplete*. Historian Joseph Tainter has shown, with both examples from the animal kingdom and historical human societies

(Rome), that high energy gain systems undergo social upheaval and ultimately collapse if they cannot maintain the energy gain that their infrastructure is built upon (Tainter 2003). The more energy required to harvest, refine and distribute energy to society, (assuming we're at maximum scale), the less will be left over for non-energy sectors. This is especially important in a society that has built its infrastructure around high-energy-return inputs (Smil 1991). Our modern situation, the energy density required for our shopping centers, hospitals, high rises, etc. is orders of magnitude higher than that of biomass and other renewables. (Smil 2006).

Net Energy and Energy Quality

In a human system, the desirability of a resource derives both from its absolute energy gain as well as from its utility to a unique sociocultural system. (Tainter 2003) Thermal energy quantity is important from a thermodynamic standpoint. However, a human society does not use or value energy based on its heat component alone. Prehistoric man would have viewed a horse as a source of meat, not as an animate converter of cropland or as a riding steed. Similarly, an ancient Yibal tribesman in Saudi Arabia would have little use for the high energy density oil bitumen just under the sands surface, but enormous use for the energy conversion capacity of a healthy horse. Today's shopping centers and hospitals could not be powered by meat calories or horsepower, but require the dense energy concentrated in fossil fuels.

Thus, energy quality is a definition dependent on the context of a society. When Watt was developing his steam engine, the heat value and liquid form of petroleum were of little use, because the new technologies of that day required wood or coal. And, unlike other mammals, humans have evolved to utilize exosomatic energy, and build and expand society around specific inanimate converters, earlier the steam engine and more recently the internal combustion engine. In this fashion, energy 'quality', as defined by an energy sources ability to perform economic or other work valued by society, can and does depart from a straight thermal assessment of the energy. Coal does not make a refrigerator work, and natural gas does not have the density to run a computer printer; these fuels must first be transformed into higher quality energy, at a thermal loss.

When assessing the quality of an alternative energy, the following factors need to be considered: energy power and density, timing, energy quality, environmental and social impacts of energy procurement and use, geographic and spatial scales, volatility, and the potential scale of the resource (energy surplus). We will now briefly discuss this first set of objective energy quality criteria. The majority of the chapter will deal with the penultimate societal energy metric; the scale of the energy surplus, and its EROI.

Energy density refers to the quantity of energy contained per unit mass or volume. The lower energy density of biomass (12–15 MJ/kg) compared to crude oil (42 MJ/kg) means that replacing the latter with the former will require a larger infrastructure (labor, capital, materials, energy) to produce an equivalent quantity of energy. (Cleveland 2007) The energy carrying molecule hydrogen, has very low energy per unit volume, creating many technical hurdles to a 'hydrogen economy', even were cheap abundant hydrogen fuel stocks available. Due to the enormous amount of geologic energy invested in their formation, fossil fuel deposits are an extraordinarily concentrated source of high-quality energy, commonly extracted with power densities of 100 to 1000 Watts/m2 for coal or hydrocarbon fields. (Cleveland 2007). This implies that very small land areas are currently used to supply enormous energy flows. In contrast, biomass energy production has densities well below 1 Watt/m2, while densities of electricity produced by water and wind are commonly below 10Watt/m2. In effect, as power dense fossil resources deplete, less power dense energy must be secured from more of the earth's surface to match the gross amount available from the concentrated high-gain sources (Smil 2006).

Bioenergy made from annual crops will also undergo unexpected volatility from periodic droughts or floods, whereas oil production can provide gasoline and its energy services continuously (or at least until a well runs dry). On a shorter time scale, the intermittency (or fraction of time that an energy source is usable to society), is low for wind and solar technologies as neither the sun nor the wind give us energy twenty four hours a day. This is potentially important with modern electricity generation systems that need to combine power generated from multiple sources and locations to supply electricity '24/7.' A derivative concept of intermittency is the dispersion over time of a source. In economics and finance, investors care greatly about the 'shape' of portfolio returns. A portfolio returning 10% consistently is much preferred to an investment that averages 15% but has periodic negative years.

In effect, investors preferences are measured by a 'risk adjusted return' which is the mean return divided by the standard deviation. Energy too, has a risk adjusted return, and constantly flowing and storable fossil fuels have built a society that depends on smooth flows of energy services. Going back to ecosystem services to procure energy may have higher standard deviations of energy availability. All natural resources show distinct geographical gradients. In the case of oil and natural gas more than 60% of known resources are in the Middle East. Just as with stored ancient sunlight, renewable energy from current sunlight (solar, wind, etc.) is geographically diffuse. This implies that significant investments (of dollars and energy) into new infrastructure will be required to concentrate, store and distribute energy over distance in order to procure useful amounts of energy services to human population centers.

Historical human energy transitions occurred when the human population was small, and had technology that was much less powerful than today. Environmental impacts associated with energy occurred locally but did not exhibit the current global impact. But the future of energy and the environment are linked, as there are numerous ecological constraints. Our future energy systems must be designed and deployed with environmental constraints that were absent from the minds of the inventors of the steam engine and internal combustion engines (Cleveland 2007).

ENERGY RETURN ON INVESTMENT – TOWARDS A CONSISTENT FRAMEWORK

Though all of the above are important factors in assessing renewable energy technologies, perhaps the most critical metric is the actual size of energy surplus freed up for society. Once an energy output becomes truly scarce – large sums of dollars won't improve its scarcity, and all the dollars in the world wont change (quickly) the demand system and energy infrastructure dependent on its energy gain. High energy gain can arise from using a resource that is of high intrinsic quality but untapped, or from technological development that allows an increase in the net energy of a previously used resource. The energy gain of mining deep coal, for example, increased greatly after Watt's engine was widely used (Wilkinson 1973). Conversely, energy gain can decline from exploiting a resource that can yield only small returns on effort under any technology, or from having depleted the most accessible reserves of a once abundant resource (Tainter 2003).

Energy Return on Investment (EROI) is an oft-confused controversial but important cousin to energy gain. EROI is basically a combined measure of how high of quality/density the original energy source is with the energy cost that the composite of harvesting technologies uses to deliver the energy to the consumptive stage. EROI is strictly a measure of energy and its 'harvesting' costs in energy terms, not the efficiency of its use or it's transformation to another energy vehicle. For example, once coal is procured out of the ground at a particular energy return, the decision, and subsequent efficiency loss to turn it into electricity or Fischer-Tropsch diesel, are both part of the consumption choices of society *after* the primary fuel is obtained.

The efficacy of EROI analysis is limited by one of its basic assumptions—that all forms of energy are fungible with a statistic determined by their thermal content (Cleveland 1992). This ignores the fact that the quality of an energy source can be the key determinant of its usefulness to society. A BTU of electricity is of higher value to society than a BTU of coal, a fact reflected by the price differential between these two energy sources as well as our willingness to convert coal into electricity at a significant energy loss. Some would argue that a technology with a low EROI should be given stronger consideration if the energy outputs have a higher quality than the energy inputs—an argument raised by Farrell et al. (2006) in support of corn ethanol which has the potential to convert coal and corn (low quality) into a liquid fuel (high quality). Cleveland (1992) has proposed a variant of EROI methodology that incorporates energy quality. Quality-adjusted economic analysis can even support sub-unity EROI energy production depending on context.

The EROI concept has been specifically used in only a small percentage of national energy analyses, but is implicit in any study that uses a form of net energy as a criterion. Recently it was used as a synthesizing concept for multiple comparisons of biofuels (Farrell et al. 2006, Hammerschlag 2006). It has been used to examine nuclear energy (Tyner et al. 1988, Kidd 2004), ethanol (Chambers et al. 1979, Pimentel 2003, Hu et al. 2004, Farrell et al. 2006, Hammerschlag 2006), other biofuels (Baines and Peet 1983, Giampietro et al. 1997, Kallivroussis et al. 2002), wood energy (Baltic and Betters 1983, Potter and Betters 1988, Gingerich and Hendrickson 1993), and other alternative energies (Crawford and Treloar 2004, Berglund and Borjesson 2006, Chui et al. 2006). Ongoing analysis continues on the EROI of various fossil fuels (Cleveland 1992, 2005, Hall, 2008).

At first blush, the calculation of EROI as the ratio of energy outputs to inputs seems straightforward. However, the concept has never expanded into common usage (Spreng 1988). Even with a recent resurgence of interest in this topic due to escalating oil prices, there is still not a widely accepted methodology for calculating either the numerator (the energy produced) or the denominator (the energy consumed) in the EROI equation. While attempting to use this important criteria to compare energy technologies, different researchers are using different methods to arrive at widely disparate notional EROI numbers, thereby diluting the policy value of this energy statistic. The ongoing heated debate over the viability of grain ethanol is a relevant example. A recent publication (Farrell et al. 2006) suggests that previous analyses of the EROI of grain ethanol are errant because of outdated data and faulty methodology. The analysis attempted to standardize previous studies and introduce modifications of the EROI methodology including measuring energy produced per unit of *petroleum* energy invested. However, because a standardized welldefined EROI formula does not exist, nor is there wide acceptance on the reasons why net energy analysis is important, the Farrell et al chapter has not ameliorated the polarization of the debate but rather heightened it (Hagens et al., 2006). At the very least, this lack of precision and consensus has negative implications for the utility of EROI analysis, in particular as a tool for decision makers. At the worst, it leaves the methodology open to manipulation by partisans in the debate over a given technology.

Furthermore, emphasis is being placed on whether or how much the energy return of a proposed technology exceeds unity, without addressing the shortfall in energy return of the segment of energy services it is trying to replace. Corn ethanol advocates and proponents spend a huge amount of resources and time honing and refining the corn-ethanol energy balance – whether it's slightly negative or slightly positive seems to be of great policy significance. At 1.5:1, which is at the high end of the latest range, corn ethanol's energy return remains an order of magnitude below the fossil energy it purports to replace (Cleveland 2001). Unless society makes large scale changes on the consumption/efficiency side, it will need to address the variance between its current energy surplus and what can be expected with the combination of lower quality fossil stocks and less energy dense renewable infrastructure in the future. Due to differences in demand, and the geographic dispersion of high energy gain renewables, there may be a variety of answers to this question at the local/regional level and at the national/global level. Since fossil fuels power a global society, global energy gain, a function of EROI times scale for all energy sources, will be of central importance in the coming decades. In the following pages, we review the various usages of EROI in the literature and place them into a consistent schematic framework. This allows comparison of the different methodologies in use by clarifying both their assumptions and their quantitative components.We then synthesize the different methodologies into a two-dimensional classification scheme with terminology for each version of EROI that will hopefully yield consistent and comparable results between studies going forward.

It is important to note that unless the energy output and input are identical types, energy extraction can still continue at an energy loss – but these joules needs to come from elsewhere in productive society. One can envision a summation of all energy technologies used globally. If we aggregate all the 'A's' (Or A+B's if we ignore environmental externalities) of all planetary energy sources, we have a sum total of energy gain for society which is able to do useful work and create human utility (beyond the sun warming us and the wind drying our laundry, and other fixed natural flows not considered in the global 500 quadrillion BTUs of annual energy use). The surplus energy of a system, or society, is what allows it to continue growth, maintenance, repair and leisure. If our energy sources required equal amounts of energy input in order to obtain an energy output, we would have no surplus energy left for other work (Gilliland 1975). If we had a very small energy surplus, we would only be able to consume at a low level.

EROI has an eventual trade-off with scale – at low scale, EROI can be very high, as the best first principles apply. At higher and higher scale, EROI eventually declines as more resources (energy and other) are needed to harvest the more difficult parts of the original resource. Indeed, analysis of the EROI of US oil and gas exploration shows that we had over 100:1 in the 1930s, when the large oil fields were discovered and put into production. By 1970 the Energy Return on Investment had declined to 30:1 and down to a range of 10–17:1 by 2000. (Cleveland 2001, Hall 2003). Anecdotally, from 2005 to 2006, the finding and production costs of the marginal barrel of oil in the US went from $15 to $35 per barrel. (Herold 2007), and offshore in the Gulf of Mexico increased from $50 to over $69 per barrel (EIA 2007). Though these are financial increases as opposed to energy, it suggests the high return oil has been found, and increasing amount of dollars (and energy) will be needed to extract the remainder.

A Framework for Analyzing EROI

Imagine the physical flows of an energy producing technology (T) e.g. a corn ethanol plant. Energy (EDin) and other various inputs ({Ik}) are taken into the plant and combined or consumed to produce energy output (EDout) as well as possibly other co-products ({Oj}) i.e. T(EDin, {Ik}) = { EDout, Oj}. In its narrowest (and least informative) form, EROI (minus 1) is similar to the economic concept of financial Return on Investment but uses energy as the currency while treating non-energy inputs as negligible. This simple definition yields EROI = EDout/EDin. EROI is rarely used in this simple form (examples being Southwide Energy Committee 1980, Gingerich and Hendrickson 1993), but

EROI statistics are frequently published regarding different technologies that ignore the energy costs associated with infrastructure and non-energy inputs (American Wind Energy Association 2006).

Non-Energy Inputs

EROI rarely conforms to the above simplistic formulation. Depending on the definition of T, the energy inputs, EDin generally do not account for additional and significant energy requirements important to the production process. This energy is embodied in the non-energy direct inputs (Odum 1983), for example the agricultural energy required to grow oilseeds for biodiesel (Hill et al. 2006). Precise calculation of the energy embodied in non-energy inputs is nearly impossible – (e.g. do we include the calories consumed by the farmer for breakfast before he goes to harvest corn? How much energy is the oil field managers expertise worth? etc.). This may be resolved either through an input-output matrix framework or by semi-arbitrarily drawing a boundary beyond which additional, (and presumably negligible), energy inputs are ignored (Spreng 1988). The latter is the accepted approach for Life Cycle Analyses (LCAs – International Standard Organization 1997). A typical EROI formulation applies an appropriate methodology to evaluate the embodied energy costs for the non-energy inputs, which are termed the indirect energy inputs. For a given production process, this should yield a specific set of coefficients, *{ãk}*, that give the per-unit indirect energy costs of *{Ik}* (e.g. MJ per tonne soybean). This gives the following version of EROI:

EROI = EDout/(EDin + _*ã*kIk). (12.1)

Some analyses arbitrarily include the indirect energy costs for certain inputs while excluding the energy cost of others, something that clearly creates difficulty of comparison between studies (Pimentel and Patzek 2005, Farrell et al. 2006). The embodied energy costs of labor in particular are difficult to define but can be a significant component of the energy cost. (Costanza 1980, Hill et al. 2006). Though energy return analysis obviously treats energy as a critical limiting variable, there are potentially numerous other limiting inputs to a production process. In addition to the direct and indirect energy requirements of an energy technology, important inputs such as land, time, and water, are difficult (some would argue impossible) to accurately reduce into energy equivalent measures. In this chapter we refer to these as *non-energy requirements* so as to distinguish them from *non-energy inputs* (which can be parsed into energy terms). Non-energy requirements can have embodied components as well (Wichelns 2001). For example, the biodiesel conversion process requires labor and water. Similarly, the oilseeds used to produce biodiesel require inputs such as land, labor, and water in addition to direct and indirect energy requirements (Pimentel et al. 1994, Pimentel 2003). The standard assumption underlying past EROI analyses is that all non-energy requirements are held constant and negligible. In a globally connected world of potentially numerous limiting inputs, energy systems analysis will benefit from a relaxing of this assumption.

The direct and indirect non-energy requirements can be handled two different ways. The first method is to identify key, potentially limiting resources and treat them completely separate from energy inputs. This would create a new indicator of efficiency for each resource tracked e.g. ERO*L*I(Land) measured in MJ/ha, or ERO*W*I(Water) measured in MJ/gallon. In particular, for non-energy requirement *X*, EROXI is given by:

EROXI = EDout/(__X,kIk) (12.2)

where ð*X*,k gives the direct and indirect per-unit requirements of *X* into Ik.

While this method increases the complexity, it also has advantages. First, it provides a metric of energy harvesting efficiency that could be included in a broader energy systems analysis. In combination with other technologies that require different array of resource inputs, this type of metric can be informative on the scaling capacity of a renewable energy portfolio. Second, this type of multicriteria approach allows for contextual assessment of a technology. Different geographic and political will be limited in their growth by different resources (Rees 1996), a Liebig's law of the minimum for economic growth (Hardin 1999). Some resources like water may be equally if not more limiting than energy (Barlow 2002). An ideal energy technology would optimize on scarce resource X (high EROXI) thus deemphasizing the return necessary on abundant resource Y (lower EROYI).

Another way to deal with non-energy primary inputs is to convert them into energy equivalents via some set of coefficients (*{X}*) for all non-energy requirements X. A justification for this is that in order for any energy procurement process to be truly sustainable, it must be able to regenerate all resources consumed (Patzek 2004). An approach adopted by Patzek (2004) and Patzek and Pimentel (2005) is to assign energy costs based on a resource's exergy (Ayres and Martinas 1995, Ayres et al. 1998), approximately defined as the ability of a system to perform work and equated with its distance from thermal equilibrium. This can also be viewed as the amount of energy necessary to reconstitute a given level of thermodynamic order.

The above set of coefficients yields the following measure for EROI:

EROI = *EDout* _ *EDin* +_ k *ãk Ik* +_ X _ k øXðX,k Ik _. (12.3)

Assuming consensus around the validity of the energy equivalents, this measure of EROI provides for complete commensurability by reducing all inputs to a single currency.

Non-Energy Outputs

Just as consideration of non-energy inputs yields a fuller, and more complex EROI statistic, so too can non-energy outputs be incorporated to provide a more complete indicator of the desirability of a process. Firstly, many technologies yield co-products in addition to a primary energy product. Most studies assume that a credit should be given for these co-products which increases the EROI by reducing the numerator for the process. Mathematically, each co-product Oj is assigned a per-unit energy equivalency coefficient (j) indicative of its value relative to the energy product.

The most straightforward method is to assign co-products an explicit energy value based on their thermal energy content (Pimental and Patzek 2005) or their exergy (Patzek and Pimentel 2005). However, co-products are seldom used for their energy content (bagasse in sugar cane ethanol being an exception). If energy is the limiting variable to be optimized, a full energy credit for dry distiller grains or milk, may be aggressive, and the EROI of a technology giving full allocation to co-products will decline as the co-products scale beyond their practical use (e.g millions of tons of DDGs). Energy values can also be assigned according to the energy that would be required to produce the most energy-efficient replacement (Hill et al. 2006). Economic value and mass are two non-energy metrics that are used to establish relative value, both of which are frequently used in life cycle analyses (International Standard Organization 1997, deBoer 2003). Once the energy equivalency coefficients have been established, the EROI formulation is modified to the following:

EROI = EDout +_ íj Oj EDin +_ ãk Ik . (12.4)

For example, when procuring biodiesel from soybeans, the soybean meal is a valuable co-product often used as a source of protein for livestock. An energy credit can be assigned to this co-product based on its actual thermal content (Pimentel and Patzek 2005), its market value (e.g. Mortimer et al. 2003), or its mass (e.g. Sheehan et al. 1998). The fact that calculated EROI can vary by a factor of 2 or more depending on allocation method gives insight that EROI, though much more so than dollars, is not a purely physical concept.

Non-Market Impacts

We have considered inputs and outputs that are currently recognized by the market system. However, many energy production processes create outputs that have social, ecological, and economic consequences external to the market. As we are all part of a planetary ecosystem, to properly include energy externalities should provide us with more accurate information of the desirability of an energy procuring technology (Hill et al. 2006). Negative externalities can include loss of topsoil erosion, water pollution, loss of animal habitat, and loss of food production capacity (Hanegraaf et al. 1998, Pimentel et al. 2002). Externalities can also be positive such as the creation of jobs and the maintenance of rural communities (Bender 1999).

As with non-energy requirements, these externalities can be incorporated into our framework in one of two ways—as separate indicators in a multicriteria framework or through conversion into energy equivalents. Thus, if topsoil is lost or nitrous oxide is emitted as part of the life cycle of the technology, we can measure EROI (*Topsoil*) or EROI(*Nox*). Studies that include such externalities have been published by the US Department of Energy (1989a, 1989b), Giampietro et al. (1997). Such measures are useful for assessing the scalability of a process within a given context by indicating what resources (e.g. waste sinks) might become limiting under increased production.

Negative externalities also can be assigned energy equivalency coefficients equal to the energy required to prevent or remediate their impacts (Cleveland and Costanza 1984, Pimental and Patzek 2005, Farrell et al. 2006). If we assume a set of externalities {Ei} with energy equivalency coefficients {_i}, then we must add into the denominator of the EROI calculation the term __iEi. Not many studies have attempted this approach, however and pursuing this strategy has the drawback of parsing important non-reducible criteria into one metric.

A UNIFYING EROI FRAMEWORK

If net energy analysis is to produce results that are clear, and comparable across studies, and be of practical use to researchers and policy-makers, it will be necessary for the methodology to become uniform and well-specified. Such standards exist in the area of life cycle analyses (International Standard Organization 1997). However, unlike LCA, it is probably not possible or even desirable that EROI be restricted to a single meaning or methodology. The different levels of energy and environmental analysis outlined above are relevant to different problems, contexts, and research objectives. *The problem heretofore has arisen when the same term is* used for methodologies with different assumptions and different goals. We propose a two-dimensional framework for EROI analyses (with accompanying terminology) that clarifies the major assumptions in an analysis. In the first dimension, we identify three distinct levels of analysis that can be distilled from the above examples. These levels differ in terms of *what* they include in their analysis.

The first level deals with only the direct inputs (energy and non-energy) and direct energy outputs. We term this *Narrow*

Boundary EROI as, while it can offer more precise EROI calculations, it is also the most superficial, restricting the analysis to simple inputs and thus missing many critical energy costs (as well as ignoring co-products). The next level, *Intermediate Boundary EROI*, involves incorporating indirect energy and non-energy inputs as well as crediting for co-products. This is the methodology used by Life Cycle Analysis to estimate the EROI of an energy technology. Intermediate Boundary EROI requires two assumptions that must be made clear: (1) What allocation method is used for the co-products (thermal content, price, mass, exergy etc.); and (2)What boundaries are used for determining indirect inputs. Finally, *Wide Boundary EROI* incorporates additional costs (and possibly benefits) for the externalities of the energy technology. Admittedly, this is the most imprecise but also the most relevant of the EROI measures in that it presents the fullest measure of the net energy available to society.

Once it has been determined what can (and should) be included in the analysis, the second dimension in our framework dictates *how* to include these inputs. We delineate three choices for handling of the non-energy requirements and externalities. They can be ignored, yielding *Basic EROI*, or converted to energy equivalents, yielding '*Total EROI*', or handled as separate components yielding '*Multi-criteria EROI*'. Note that while the grid is 3×3, it yields only 8 meaningful formulations. The different levels of analyses are nested hierarchically. The computation of a wider boundary EROI for an energy production process should easily yield all other forms of EROI found below it. That is to say, the necessary data will have been compiled and it is merely a decision of which components to include in the calculation. Similarly, a Total EROI calculation will use the same data set as a Multi-criteria EROI with the addition of energy equivalency coefficients. This means that more comprehensive studies should yield results at least partially comparable with less comprehensive studies as seen in a meta-study of ethanol by Farrell et al. (2006).

Liebig's Law, Multi-Criteria Analysis, and Energy from Biofuels

Though it is becoming apparent that energy will be a limiting variable for society going forward, it is easy to envision other equally limiting variables as the planetary population increases its demand on ecosystems. Water, land, and carbon sinks are only three examples of inputs and impacts of renewable energy production that could limit the potential of a technology (Giampietro et al. 1997, Hagens et al. 2006, Hill et al. 2006). These should be included explicitly in a net energy analysis or else their cost in terms of energy should be estimated.

Liebig's Law of the minimum states that the production of a good or resource is limited by its least available input. In layman's terms something is only as good as its weakest link. This form of ecological stoichiometry will loom large in the procurement of energy alternatives to fossil fuels. Water, land, soil, greenhouse gas emissions, and specific fossil inputs themselves will potentially limit scaling of alternative energy.

Though EROI is generally measured as the ratio of the gross energy return to the amount of energy invested, it has been argued this can give a false indicator of the desirability of a process due to the increasing cost of non-energy requirements as EROI approaches 1. Following Giampietro et al. (1997), let = EROI/ (EROI – 1) be the ratio of gross to net energy produced. equals the amount of energy production required to yield 1 MJ of net energy. From an energy perspective, all costs have been covered. However, for non-energy requirements the perspective and the implications, change.

Let EROXI be the energy return for 1 unit of non-energy requirement X. Then 1/EROXI is the number of units of X required for 1 MJ gross energy production. From the above, it is easily seen that /EROXI units of X are required, or more generally, the net energy yielded per unit of X is equal to EROXI/. Since increases non-linearly (approaching infinity) as EROI approaches 1, a relatively small change in EROI can produce a large decrease in the 'net EROI' for non-energy requirements. For energy production processes with significant non-energy requirements such as biofuels, this suggests a low EROI can imply strong limitations on their ability to be scaled up (Giampietro et al. 1997, Hill et al. 2006).

If we assume the Intermediate Boundary EROI for non-cellulosic ethanol from corn is in the neighborhood of 1.34 (Farrell et al. 2006), this implies net energy of .34 for every 1 unit of energy input. The corn-based ethanol Energy Return on Land Invested (EROLI) = 11,633 MJ/ha gross energy production (equivalent to 3475 l per hectare). However, the net energy per unit of land is only 2,908 MJ/ha. At 2004 levels of gasoline consumption for the United States, this is equivalent to consuming the net energy production of 42 ha of cropland per second. If the EROI of ethanol is reduced to 1.2, a decrease of only 10%, the net return on land decreases by 33% while the amount of land required to achieve this same net yield increases by 50%. Conversely, an oil well requires equipment access, roads, etc. but pulls its bounty out of a comparatively small land area. This contrast has significant implications for the potential scale of biofuel production (Giampietro et al. 1997). In effect, due to significant power density differentials, replacing energy-dense liquid fuels from crude oil with less power dense biomass fuels will utilize 1,000- to 10,000-fold increases in land area relative to our existing energy infrastructure (Cleveland 2007).

Though land is one limiting factor, water may be another. In a forthcoming paper, we use Multicriteria EROI analysis to define and quantify the EROWI (Energy Return onWater Invested) for various energy production technologies. Since water and energy may both be limiting, we care about the 'Net EROWI', which is a combined measure of EROI and EROWI for each technology. With the exception of wind and solar which use water only in indirect inputs, the 'Net EROWI' of biofuels are one to two orders of magnitude lower than conventional fossil fuels.We also determined that approximately 2/3 of the world population (by country) will have limitations on bioenergy production by 2025, due to other demands for water (Mulder et al. *In press*).

Nitrogen, a byproduct of natural gas via ammonia, is essential to a plant's ability to develop proteins and enzymes in order to mature. The importance of nitrogen fertilizers to U.S. agriculture, particularly corn and wheat, is evidenced by its accelerated use over the last 50 years. From 1960 to 2005, annual use of chemical nitrogen fertilizers in U.S. agriculture increased from 2.7 million nutrient tons to 12.3 million nutrient tons (Huang 2007). This increase is considered to be one of the main factors behind increased

U.S. crop yields and the high quality of U.S. agricultural products (Huang 2007). Furthermore, biofuels, especially the ethanols, require large amounts of natural gas for pesticides, seedstock and primary electricity to concentrate the ethanol. In areas that have natural gas fired electricity plants (as opposed to coal), fully 84% of the energy inputs into corn ethanol are from natural gas (the nitrogen, a portion of the pesticides, and the electricity). (Shapouri 2002). Ethanol proponents, other than optimizing 'dollars' (making money), are presuming that 'domestically produced vehicle fuel' is the sole item in short supply. Were the math on corn ethanol somehow scalable to 30% of our national gasoline consumption, in addition to land and water, we would use *more than the entire yearly amount of natural gas currently used for home heating* as an input.

Though many biofuel studies imply that fertilizer, and therefore natural gas, are more abundant and cheaper than petroleum, we are actually on a 'natural gas treadmill' in North America and low prices are being kept down only by 2 consecutive mild winters and summers with no hurricanes. In 1995 the average new gas well in North America took 10 years to deplete. A new gas well in 2007 takes under 10 months. More and more drilling of new gas wells is necessary just to stay at constant levels of production. US production peaked in 1973 followed by another peak in 2001.

The second peak required 370% more wells to produce the same amount of gas. Furthermore, the energy/$ effort on Canadian natural gas production implies a decline in EROI from 40:1 to 15:1 from 2000 to 2006, with an extrapolated energy break even year circa 2014. (CAPP 2007, methodology Hall and Lavine 1979). The falling EROI makes it impossible for natural gas production to maintain both low costs and current levels of production. When US oil peaked in 1970, we made up our oil demand shortfall by imports. Natural gas can also be imported (as LNG), but it must first be liquefied at a high dollar and energy cost. It requires over 30% of its BTU content to be transported overseas – another energy loss. In this sense, studies that show energy use on petroleum invested are perhaps overlooking natural gas as a limiting input.

So corn ethanol, and other biofuels requiring both natural gas for fertilizers and pesticides, as well as for electricity to steam the ethanol solution, are essentially turning 3 scarce resources: water, land, and natural gas, into liquid fuels, at an energy gain an order of magnitude lower than what societal infrastructure is currently adapted to. What will the strategy and metrics to measure it become when natural gas too, is recognized as limiting input?

CONCLUSION

At some point in the near future, those reading this chapter will witness a forced change from the fossil fuel mix that has powered society smoothly for decades. In a perfect world, all information about externalities and an accurate balance sheet of the size and quality of our resources would be available to decision-makers. In reality however, accurate information about the reliability of upcoming resource flows is opaque beyond a few months. Only 6% of the worlds (stated) oil reserves are owned by public companies subject to SEC requirements, leaving the NOCs and private companies each individually knowing only their own share of the oil pie. It is unlikely the market will respond in time once critical limiting variables to society become apparent.Unfortunately, this cannot be empirically proven until after the fact. To have a framework in hand that anticipates such problems is a first but important step.

New energy technologies require enormous capital investments and significant lead time as well as well-defined research and planning. Aggregating decisions surrounding alternative energy technologies and infrastructure will be both difficult and time sensitive. As a growing population attempts to replace this era of easy energy with alternatives, net energy analysis will reassert its importance in academic and policy discussions. Alongside ecological economics, it is one of the few methods we can use to attempt to measure our 'real' wealth and its costs. As such, it will be advantageous to adhere to a framework that is consistent among users and attempts to evaluate correctly the complex inputs and outputs in energy analysis in ways that are meaningful. Accounting for the subtle and intricate details in net energy analysis is difficult. However, in a growing world constrained by both energy and increasingly by environmental concerns, adherence to a common framework will be essential for policy-makers to accurately assess alternatives and speak a common language.

Perhaps the biggest misconception of net energy analysis, particularly in its most popular usage referring to corn ethanol, is the comparison on whether or not something is energy positive – this myopic focus on the absolute, ignores the much larger question of relative comparisons – what happens to society when we switch to a lower energy gain system? While net energy analysis outcomes will not guide our path towards sustainable energy with the precision of a surgical tool, they are quite effective as a blunt instrument, helping us to discard energy dead-ends that would be wasteful uses of our remaining high quality fossil sources and perhaps equally as important, our time. Ultimately when faced with resource depletion and a transition of stock-based to flow-based resources, EROI will function best as an allocation device, marrying our demand structure with our supply structure, thus guiding our high quality energy capital into the best long term energy investments. Finally, analysts and policymakers may use net energy analysis not only to compare the merits of proposed new energy technologies, but also as a roadmap for possible limitations on demand, if global energy systems analysis points to declines in net energy not adequately offset by conservation, technology or efficiency. A framework like the one presented above, may also be useful for analyses involving limiting inputs in addition to energy.

10

Bio-ethanol Production in Brazil

In this chapter the history and origin of the Brazilian program for bioethanol production (ProA´lcool) from sugarcane (*Saccharum* sp.) are described. Sugarcane today covers approximately 7Mha, with 357 operating cane mills/ distilleries. The mean cane yield is 76.6Mgha"1 and almost half of the national production is dedicated to ethanol production, the remainder to sugar and other comestibles. The mean ethanol yield is 6280 L ha"1. An evaluation of the environmental impact of this program is reported, with especial emphasis on a detailed and transparent assessment of the energy balance and greenhouse gas (CO2, N2O, CH4) emissions. It was estimated that the energy balance (the ratio of total energy in the biofuel to fossil energy invested in its manufacture) was approximately 9.0, and the use of ethanol to fuel the average Brazilian car powered by a FlexFuel motor would incur an economy of 73% in greenhouse gas emissions per km travelled compared to the Brazilian gasohol. Other aspects of the environmental impact are not so positive. Air pollution due to pre-harvest burning of cane can have serious effects on children and elderly people when conditions are especially dry.

However, cane burning is gradually being phased out with the introduction of mechanised greencane harvesting. Water pollution was a serious problem early in the program but the return of distillery waste (vinasse) and other effluents to the field have now virtually eliminated this problem. Soil erosion can be severe on sloping land on susceptible soils but with the introduction of no-till techniques and green-cane harvesting the situation is slowly improving. The distribution of the sugar cane industry shows that reserves of biodiversity such as Amazˆonia are not threatened by the expansion of the program and while there may be no great advantages of the program for rural poor, the idea that it will create food shortages is belied by the huge area of Brazil compared to the area of cane planted. Working conditions for the cane cutters are severe, almost inhuman, but there is no shortage of men (and women) to perform this task as wages and employment benefits are considerably more favourable than for the majority of rural workers. The future will bring expansion of the industry with increased efficiency, more mechanisation of the harvest, lower environmental impact along with a reduction in the number of unskilled workers employed and an increase in wages for the more skilled. This biofuel program will not only be of considerable economic and environmental benefit to Brazil, but also will play a small but significant global role in the mitigation of greenhouse gas emissions from motor vehicles to the atmosphere of this planet.

HISTORICAL INTRODUCTION

The present large Brazilian program for bioethanol production is historically derived from the introduction of the sugarcane plant (*Saccharum officinarum*) from the island of Madeira by the Portuguese colonising expedition of 1532 (Machado et al., 1987). At that time Brazil was a Portuguese colony in South America, and its first economic cycle was based only upon natural resources such as brazilwood (*Caesalpinia echinata*), gold and precious stones.

Soon after the exploration of the interior of the country, sugar-cane became the first large-scale plantation crop, and depended on the labour of slaves in the newly-opened wilderness. Until the end of 19thCentury, cultures such as rubber (*Hevea brasiliensis*) and coffee (*Coffea arabica*) occasionally eclipsed its economic importance. In the colonial period, there was a productive rural structure of traditionally midto- large-size estates that contributed to populate the interior of the country. The edaphoclimatic conditions in S˜ao Paulo and Rio de Janeiro States in the southeast, and Pernambuco State in the northeast, favoured the spread of this crop in these regions. After the abolition of slavery in 1883, the supply of cheap labour to cut cane was initially maintained by the arrival of European immigrants. Consequently, the processing units for sugar production, and later the attached bioethanol distilleries, were closely related to a traditional oligarchy with a resolute and lasting political influence on the country's affairs.

The first trials on the use of ethanol blends in petrol engines took place in the early years of Getulio Vargas dictatorship, soon after the foundation in 1933 of The Sugar and Alcohol Institute (Instituto do Ac¸u´car e do A´ lcool, IAA). Extensive use of anhydrous bioethanol was attempted during the course of Word War II in order to save oil imports. Later on in 1953, during the democratically-elected second Vargas presidency, the major national oil company, Petrobras, was founded to promote fuel production and industrial development.

When the Oil Crisis of 1973 hit the international fuel supplies, Brazil was importing 72% of its crude oil, and was almost completely dependent on petroleum derivatives for the transport sector. Oil import expenses rose from US$ 600 million that year up to US$ 2.6 billion in 1974. In this period the annual balance of payments changed from a small surplus to a deficit of US$ 4.7 billion. It was against this background that in 1975 the military dictatorship created the National Alcohol Programme (PROA´ LCOOL), with the aim of moving towards the introduction of engines fuelled solely by hydrated ethanol. The first automobiles running on ethanol and other bio-fuels were developed at Centre for Aerospace Technology (Centro T´ecnico Aeroespacial, CTA), a Research Centre of the Brazilian Air Force, located at S˜ao Jos´e dos Campos, S˜ao Paulo State. The motor vehicle industry principally led by the multinational companies Volkswagen, Ford, Fiat and General Motors started large-scale production and new parts and materials were soon developed to resist corrosion and solve the problem of starting the engines from cold. Ethanol production was 500,000 litres per year in 1975 at the beginning of PROA´ LCOOL (and reached 3.4 billion litres only five years later – TCU, 1990). A complete and distinct program of tax and investments was brought out to support PROA´ LCOOL, for the industrial sector of new distilleries and enlargements, for sugar-cane farming and for final ethanol consumption. Up to 1990 the investment amounted to more than US$ 7 billion, with almost US$

4 billion of public resources. After 1990 no more direct subsidies were supplied by the government but as gasoline was taxed at a much higher rate, cars and other light vehicles were cheaper to run on ethanol and sales from 1983 until 1989 of light vehicles running this fuel outstripped gasoline vehicles. The main problem with the program was that in the late 1980s and through the 1990s crude oil prices declined to below US$ 20 a barrel. Petrobras became very antagonistic to the ethanol program as gasoline was being substituted by ethanol. As a consequence, in order to provide the home market with sufficient diesel and naphtha the company was left with excess gasoline that had to be sold at low prices on the international market. Added to this there were several crises, caused by high international sugar prices and low rainfall that lowered ethanol production, and in some years (1989 and 1990) there were huge queues for ethanol at the gas stations and car buyers lost faith in relying on this biofuel.

It can been seen from the production figures that in 1988 (when 95% of cars being manufactured were equipped with alcohol engines), hydrated ethanol reached 9.5 billion litres but then varied between 8.7 and 10.7 billion litres until 1999 (9.25 billion litres). By this time very few ethanol-powered cars were being produced and much of the ageing fleet had left the roads, such that in 2000 production fell to less than 7 billion litres, reached a low of just under 5 billion litres in 2001 and only exceeded 7 billion litres again after 2005.

However, the government could not let the program die, as apart from the pressure from the powerful cane planters lobby, more than 700,000 desperately-needed jobs had been created in the rural sector (TCU, 1990). For this reason in 2001, a law was passed making obligatory to add between 20 and 24% of anhydrous ethanol to all gasoline (Federal Law No. 10,203 of 22nd February). Historically, all over the world tetraethyl lead was added to gasoline to avoid spontaneous combustion before (spark) ignition. This enhancement of octane rating could also be achieved with the addition of ethanol. In fact this was well known, and published in Scientific American a few years before Thomas Midgely in the USA synthesised tetra-ethyl lead in 1922 (see Kovarik, 2005). The change from leaded gasoline to gasohol therefore was perceived to have a beneficial effect on air quality, especially in urban areas, and of course was extremely popular with the sugarcane industry.

However, the great leap forward for Brazilian bioethanol has just begun with the invention and production of ethanol/gasoline FlexFuel Otto cycle engines. Flex-fuel engines were first released in March 2003, a joint project of Volkswagen and Bosch. The compression ration of the engines is between 10:1 and 12.5:1 intermediate between that for gasoline (9–10:1) and ethanol (13–14:1). A carburettor control unit receives two basic signals. A conductivity detector informs the composition of the fuel in the tank and an oxygen probe analyses the concentration of this element in the exhaust vapour. The control unit electronically regulates the air-fuel mixture in order to reach the right stoichiometric rate for optimal burning of any ethanol/gasoline combination. This innovation has coincided with the increase of international crude oil prices, which since 2000 have risen above US$ 30 to between US$ 50 and US$ 100 today.

Until end of July 2006, 2 million FlexFuel powered vehicles were sold and from August 2006 to May 2007 another 1.3 million, totalling 3.3 million (ANFAVEA, 2007). From January to May 2007, 67% of all Otto cycle vehicles sold were Flexfuel, the remainder running on gasohol (20–24% anhydrous ethanol). In June 2007 this proportion reached 89.7%.

THE SUGARCANE CROP IN BRAZIL

The Situation Today

With the great international interest in bio-ethanol, the area planted to sugar cane is rapidly expanding. For the 2007 season it is estimated that 7.8Mha of sugarcane will be planted, an increase of 9.9% over 2006. More than half of the area (55%) planted to cane in Brazil is in the state of S˜ao Paulo, and this area increased by 10% over the last year. While 1.2 Mha was planted in north eastern states, this area has not increased appreciably, and the largest proportional increases have been in the Cerrado (central western savanna) region with an increase of 35% in Mato Grosso do Sul, 20% in Minas Gerais and in the southern state of Paran´a (26.5%). S˜ao Paulo, and these three states where the area is expanding most rapidly, account for 73.4% of the planted area, and as yields are well above the national average, these states contribute 77.5% of national cane production.

Sugar and Ethanol Production

From sugarcane, Brazil produces sugar, hydrous ethanol (5% water) for use in motors adapted for this fuel, anhydrous ethanol (<0.5% water) for mixing with gasoline, and other products such as the alcoholic beverage cachac a and various other products such as molasses and "rapadura" (a traditional sweet cake). Flex Fuel motors can function on any mixture of hydrous or anhydrous ethanol with gasoline.

The data from the Ministry of Agriculture (MAPA, 2007) available for the 2005/2006 harvest, only includes sugar and ethanol. Of the "total recovered sugar" exactly 50% was used to produce refined sugar and 50% for ethanol. For the two types of ethanol 49% was anhydrous and 51% hydrated. Predictions were made recently (31st May 2007) by the Ministry of Agriculture that total cane production form the 2007/2008 seasonwould be 582 millionMg, of which 44.8% will be used to produce ethanol fuel, 43.9% for refined sugar and the remaining 11.3% for other products (UOL Economia, 2007). In the 2006/2007 season the production of ethanol was approximately 17.5 billion litres, and for the next year it is estimated at 20.0 billion litres. The present yield of hydrated ethanol perMg of cane (fresh weight) is estimated to be 82.0 L (MAPA, 2007) which is close to the value of 85.4 L given by Macedo (1998) for the State of S˜ao Paulo. Thus using the mean national value and estimated yield for 2006/2007 of 76.6Mg cane ha"1 (IBGE, 2007), one ha of sugarcane produced 6281 L of ethanol ha"1.

The Crop Cycle

In S˜ao Paulo, and in more productive areas of other states, general practice is to plant cane every 6 years. The first (plant) crop is harvested approximately 18 months after planting, and then there are four subsequent ratoon crops which are harvested at 12-monthly intervals (Macedo, 1998). The land generally lies fallow for the 6 months until the next planting, although occasionally a 'break crop' of groundnut or soybean is grown during this period. This practice is not common as few plantation owners have access to the necessary machinery for planting and harvesting crops other than sugarcane.

Land Preparation (Tillage)

Tillage prior to planting is usually intense, with sub-soiling followed by two or three passes with a heavy disc plough before harrowing and the subsequent formation of the furrows. As cane is planted from setts (lengths of cane stems), the furrows are 20–30 cm wide and approximately the same depth. With thewidespread introduction of zero tillage (ZT) in themechanised production of grains in Brazil, this practice has recently been adapted for the sugarcane crop. All existing weeds and cane regrowth are treated with herbicide and the only mechanical operation is the furrow making. As the furrows are comparatively wide, a certain proportion of the soil is disturbed, but as spacing is also wide (usually 1.4–1.5m between rows), this means that 80% or less of the soil surface is not tilled. This should lead to better maintenance of soil structure and aggregate integrity and probably favour soil organic matter accumulation ("C sequestration" – Six et al., 2000), but as this technique has only recently been introduced there do not yet appear to be any studies on the impact of the introduction of ZT on soil carbon stocks. Virtually all planting is from setts usually produced on-farm and approximately 12Mg of setts are required per ha.

Fertilisation

At planting the setts are covered with filtercake (from the large filters used to remove suspended material from the cane juice) at between 10 and 20Mgha"1. Typical nutrient content of this material is an addition of 10Mg per ha would amount to an input of 63 kg N, 77 kg P, 15 kg K, 100 kg Ca and 49 kg Mg. In addition best practice (Macedo, 1998) is to add 500 kg of 4-24-24 fertiliser hence adding 20 kg N, 120 kg P2O5 and 120 kg K2Oha"1. Many agronomists and others have reported that there is very rarely a response of the plant crop to N fertiliser (Azeredo et al., 1986). Ratoon crops do usually respond to N fertiliser but rarely more than 100 kgNha"1 are applied. Assuming an application of 20 kgNha"1 at planting and 80 kgNha"1 for each of the 4 ratoon crops spread over a 6 year cycle, the annual mean application becomes 56.7 kgNha"1. Nearly all other cane-producing countries of the world utilise at least 150 kgNha"1 yr"1, and most approximately 200 kgNha"1.

The reason for much lower use of N fertiliser on cane in Brazil seems to be partially that Brazilian cane varieties, which were first bred in soils of low N fertility, are able to obtain significant inputs of N from association with N2-fixing bacteria (Lima et al., 1987; Urquiaga et al., 1992; Boddey et al., 2001). There is a considerable amount of literature on this controversial subject and readers are referred to reviews of James (2000), Baldani et al. (2002) and Boddey et al. (2003). However, there is no question that sugarcane has been grown for many decades, often centuries, in many regions of Brazil with no apparent long-term decline in yields or soil fertility, even though it is estimated that more N is removed by export of cane to the mill and trash burning than is added as N fertiliser (Boddey, 1995). The N2- fixing bacteria that infect the interior of the plant tissues (endophytic diazotrophs) are generally thought to be responsible for most of the input from BNF (Baldani et al., 1997; James, 2000). Sugar cane plants have been found to be infected with very significant numbers of such diazotrophs in other countries such as Australia (Li and Macrae, 1992), Mexico (Mu˜noz-Rojas and Caballero-Mellado, 2003) and India (Muthukumarasamy et al., 1999, 2002). However, attempts to prove that BNF inputs to sugarcane are of agronomic significance in countries other than Brazil have not been successful (Biggs et al., 2002; Hoefsloot et al., 2005).

Cane Harvesting

Before the 1940s, pre-harvest burning of sugarcane in Brazil was virtually unknown. However, subsequently with the increasing price of labour, pre-harvest burning became almost universal until a few years ago. Until recently, virtually all cane was manually harvested, and one man in one day can manually harvest almost three times as much burned cane as unburned cane. It was the introduction of mechanical harvesting which facilitated the return to trash conservation. A strong lobby of environmentalists, who were especially active in the state of S˜ao Paulo, claimed that there were serious human health dangers (respiratory problems) with the annual cane burning. This has led to legislation in this State that mandates that all pre-harvest burning of cane must be phased out by the year 2022. Only on land that has greater than a 12% slope, where machine harvesting is non-viable, will burning be allowed until 2032. Today approximately 20% of sugarcane is not subject to pre-harvest burning (green cane harvesting), and most of this area is in S˜ao Paulo (Coelho, 2005).

As one cane harvesting machine can replace 80–100 men, in a country with large pockets of acute rural under- and unemployment, there are considerable negative social consequences of this change in practice. However, from an agronomic point of view the conservation of trash has considerable benefits. Our team at Embrapa Agrobiologia recently completed a 16-year study on the effects of pre-harvest burning versus trash conservation on cane productivity and soil organic matter content (Resende et al., 2006). The study was conducted in Pernambuco, some 100 km from the coast, where rainfall is often sub-optimal for cane production. The results showed clearly that the conservation of trash had most benefit in dry years, and that over the whole 16 years, trash conservation increased cane yields by 25% from a mean of 46 to 58Mgha"1. In the same study it was found that soil carbon stocks to 20 or 60 cm depth were not significantly affected by trash conservation. There was a tendency for the unburned plots to have accumulated annually a mean of 90 kg (0–60 cm) to 150 kg C ha"1 (0–20 cm).

Relatively short-term studies at two sites in S˜ao Paulo, both close to the city of Ribeir˜ao Preto, have been reported, one on an Oxisol (Hapludox) and the other on an Entisol (Quartzipsamment) (Campos, 2004). The accumulation of trash in the unburned cane fields reached respectively 4.5 and 3.6Mg dry matter ha"1 after 4 years without burning. The author concluded that an annual rate could be calculated for C accumulation from these data, but the decomposable fraction was probably achieving steady state by this time. He also observed an increase of approximately 1MgC ha"1 yr"1 in the soil during this period, but there was no replanting of the cane during this period. Based on these data Cerri et al. (2004) and Mello et al. (2006) suggest that the change from pre-harvest burning to trash conservation would promote a mean soil

C accumulation of 1.62MgCha"1 yr"1. As explained above, when cane is replanted, heavy tillage and deep plowing are used which lead to large mineralization losses of soil organic matter. For this reason, the difference between the SOC stocks under burned and green cane are not likely to reach even 1MgCha"1 yr"1 over the long-term (Boddey et al., 2006). On the other hand, the values from the EMBRAPA Agrobiologia experiment in Pernambuco may be lower than a mean for S˜ao Paulo, in that this area of Pernambuco often has years with low yields due to lack of rainfall, and mean yields for the region are only about 60% of those for the State of S˜ao Paulo.

ENVIRONMENTAL IMPACT

To consider the environmental impact of the production of bioethanol from sugarcane, and the expansion of this activity, the following items are considered: A. Global impact: The energy balance of the bio-ethanol production and the impact on greenhouse gas emissions;

B. Local and Regional impact: Atmospheric and water pollution, and soil erosion.

ENERGY BALANCE

Every biofuel requires as least some input of fossil fuel in its manufacture and distribution. Starting with the agricultural operations there are inputs of diesel fuel for ploughing, transporting seeds etc, then for harvesting, factory processing and fuel distribution. To calculate the balance for bioethanol produced from sugarcane in Brazil, we used the most recent available data for all inputs and divided the energy inputs into the following categories:

A. Agricultural operations

B. Transport of cane to mill/distillery and of raw materials from suppliers

C. Factory/distillery operations

Different authors have used different units (often not metric but "Imperial") or for expressing areas, crop yields, fertilisers and units of fuel production. In this chapter we use only SI (metric units) and express energy as joules (MJ or GJ). We have calculated all energy inputs and outputs on a per ha basis. The justification for this is that for agricultural operations the energy used in tillage operations, seeding and harvesting, are usually very similar on a per ha basis regardless of crop yield. Many authors do not include energy required to build factories and distilleries, and to fabricate farm vehicles and transportation equipment (Sheehan et al., 1998; Shapouri et al., 2002). However, the relevant ISO standard (ISSO 14040–ISO, 2005) for Life Cycle Assessment studies clearly states that "manufacture, maintenance and decommissioning of capital equipment" should be taken into consideration.

FUEL FOR AGRICULTURAL OPERATIONS

As mentioned above normal good practise is to plant cane every six years, and subsequently harvest the plant crop and four subsequent ratoon crops. In areas where the plantations are replanted at longer intervals, diesel fuel consumption per year will be lower on an annual basis. For planting cane it is estimated that approximately 99 L of diesel are used per ha, and as no tillage operations are involved, maintenance of the ratoon crops requires far less fuel (<9Lha"1). Thus the weighted mean annual diesel consumption is 22.3 L ha"1, which at 47.7MJ L"1 (11.414 Mcal L"1 – Pimentel, 1980) gives total input of 1063 MJ ha"1 yr"1.

AGRICULTURAL INPUTS

Apart from the diesel fuel energy input computed above, the other fossil energy agricultural inputs are derived from, human labour, industrial fertilisers, seed material, pesticides and the energy utilised to manufacture and maintain the agricultural implements.

Manual labour: The largest labour input is in the harvesting of the cane that is still burned and then cut by hand in approximately 80% of the area in Brazil (Coelho, 2005). A basic "Tarefa" (literal translation = task) for one man to cut cane is 6Mg per day, and while nearly all workers cut more than 1 tarefa per day, the ratio of manual energy to cut 1 ton of cane is the same. Even if very conservatively we assume that it takes one man 8 hours to cut 6Mg of cane then for each ha in Brazil which yields a mean of 76.6Mg fresh cane ha"1, it requires 76.6/6 × 8 = 102.1 man hours ha"1 harvest"1. As there are 5 harvests in 6 years this becomes 85.1 man hours ha"1 yr"1. Considering that apart from cane cutting there is manual planting, weeding and many other minor tasks, the estimate of Pimentel and Patzek (2007) of 128 h ha"1 yr"1 does not seem unreasonable.

Most authors who calculate energy balance for biofuel crops do not count any energy input for human labour. However, as each individual consumes fossil energy to survive and work it seems logical to include their fossil energy consumption as an input to the cane/ethanol production system. Giampietro and Pimentel (1990) estimated that in poor/rural societies such energy inputs range from 25.1 to 62.7MJ day"1 (6000–15000 kcal day"1). Utilising the higher value and assuming that all this energy is utilised in field labour, 1 man hour is equivalent to 62.7/8 = 7.84 MJ. The total energy invested in manual labour thus becomes 1003.5MJ ha"1 yr"1.

FERTILISERS

As calculated above it is estimated that mean N fertiliser inputs to cane are 56.7 kgNha"1 yr"1. Smil (2001) shows that N fertiliser production has greatly improved in energetic efficiency over the past 50 years from >80 GJMg"1 NH3 before 1955 to 27 GJMg"1 NH3 in the most efficient plants operating in the late 1990s. The mean value given by Lægreid et al. (1999, p. 204) is 54 MJ kgN"1 for urea production in plants operating in 1999 and this value was adopted giving an overall fossil energy cost of 3062 MJ ha"1 yr"1. These same authors give values of 3.19 and 5.89 MJ kg for P and K, respectively. From our mean annual estimates of 16 and 83 kg ha"1 of P and K applied, respectively, the energy costs become 51 and 489 MJ ha"1 yr"1 for these two fertilisers, respectively. The other major soil amendment is lime, in that virtually all soils in Brazil used for sugar cane are acidic. Macedo (1998) estimates that as every replanting of the cane (every 6 year) a mean of 2Mg ha"1 of lime area added. This value is appropriate for opening up new land (mainly degraded pastures), which have not been limed for several years, but is higher than would be used for a plantation that has been operating for few decades or more.

However, using this value of 2Mgha"1 (367 kg lime ha"1 yr"1), and an energy cost of 1.31MJ kg for lime manufacture (Pimentel, 1980) we estimate a total annual fossil energy cost of 479 MJ ha"1.

Pesticides

Brazil has probably the largest program (in terms of land area) of any country in the world for biological control of insect pests and this is precisely on the sugarcane crop. The most widespread insect pest of cane is the sugar cane borer (*Diatraea saccharalis*), which can cause serious damage to cane in all regions of Brazil, damage being mainly secondary due to invasion of the tunnels bored in the stems by this pest by fungi. Control was at first by species of flies native to Brazil (*Metagonistlum minense* and *Paratheresia claripalpis*) but now almost universally the introduced wasp (*Cotesia flavipes*) is used, all of which lay their eggs in the larvae of the stem borer (Botelho, 1992). The *C. flavipes* gives the most effective control and is widely used in many cane growing areas by releasing hundreds of thousands of flies//wasps into the fields. At present over 1Mha of cane are treated with *C. flavipes* to control Diatraea and more than 20 companies are engaged in producing this control agent and the number is growing (Sene Pinto, 2007).

The other main pests are restricted to the northeast region of Brazil and down the coast as far south as Rio de Janeiro, and are the root spittle bug (*Mahanarva fimbriolata),* which sucks root sap, and the froghopper (*Mahanarva posticata*) which sucks leaf sap. The main damage to the plant is due to toxins injected into the plant phloem at the time of penetration of the insect stylet. Control is by spraying the fields with the fungus *Metarizium anisopliae,* which parasitises the exoskeleton of the sap-sucking pests. This fungus is generally produced on-farm by inoculating sterile boiled rice. A suspension of the fungus/rice (which breaks up into a slurry when vigorously agitated with water) is spayed onto the leaves. This control program is applied on approximately 600.000 ha of cane mainly in the north-eastern region (Sene Pinto, 2007).

There are many other insect pests, but none that have the potential to cause such widespread damage as the stem borer, spittlebug or froghopper. These minor pests are generally controlled by insecticides, and in consequence Brazil's use of insecticides is far lower than that used on other crops such as citrus, coffee or soybean. According to the National Association of Pesticide Manufacturers, in 2006 a total of 1700Mg of insecticide (active ingredient – a.i.) of all insecticides were used on 7.1Mha of cane, a mean of 0.24 kg ha"1 (SINDAG, 2007). The same source shows data that only 1Mg (!!) of fungicide was sold to cane producers. However, as especially in recent years, weed control has become almost universally chemical (mainly glyphosate), herbicide sales for 2006 were 22,851 Mg, a mean of 3.2 kg a.i. ha"1. According to Pimentel (1980) the fossil energy cost of the insecticides most utilised on sugarcane (Carbofuran, Diuron and Endosulfan) is approximately 364 MJ kg a.i."1 (87Mcal kg"1). The herbicide glyphosate is cited as having an energy cost of 452 MJ kg a.i."1 (108 Mcal kg"1).We calculate therefore the energy fossil energy inputs for insecticide and herbicide are, respectively, 87 and 1445 MJ ha"1.

Planting Material

Virtually all sugarcane is planted from setts (stem pieces), and 12 Mg of setts are required per ha every 6 years (Macedo, 1998). The means 2Mg per year out of 76.6Mg ha"1, or 2.6%. As the agricultural operations for sett production are the same as for the rest of the cane plantation, the energy input for seeds is regarded as 2.6% of the total agricultural fossil energy input (TAFEI). Thus the fossil energy for seed production = (0.026 × (TAFEI* × 100/ (100–2.6)), where TAFEI* is the TAFEI excepting the energy input in the setts. The fossil energy for sett production was estimated to be 252.2GJ ha"1 yr"1. Irrigation

Cane is only planted in regions where there is usually sufficient annual rainfall for the crop. Only a very small proportion of the sugarcane area in Brazil is irrigated, but in almost all cases the distillery waste (vinhasse) is applied to the fields. Between 10 and 12 L of vinasse are produced per L of ethanol and the return of this to the fields is valuable source of nutrients, especially potassium. Usually approximately 80m3 of vinasse are applied per ha and Resende et al. (2006) calculated that this adds 23 kg N, 8 kg P, 93 kgK and 35 kg S ha"1. The vinasse is mixed with waste water used to wash the cane before grinding, typically giving a total volume of diluted vinasse of 160–200m3 ha″1 yr″1. According to Dr Rog´erio P. Xavier (Usina Itamaraty, Mato Grosso) a diesel pump of 125 HP is requires 2 h to irrigate 1 ha with this volume of diluted vinhasse, which incurs a fossil energy input of 656 MJ ha″1.

AGRICULTURAL MACHINERY

Macedo et al. (2003) gives the density of utilisation of equipment for tractors and harvesters of 41.8 kg ha″1. This would mean for a 20,000 ha plantation there would be 836Mg of machinery, which certainly does not appear to be an underestimate. For implements towed by tractors he gives 12.4 kg ha″1, and for transport vehicles to haul cane etc., he gives 82.4 kg ha″1.

He cites Pimentel (1980) for the methodology used to calculate energy input from these data as follows:

(a) The energy incorporated in the material (steel, rubber for tyres etc.) and for the fabrication, repairs and maintenance are considered. The energy incorporated in this case is essentially in the steel and tyres. The energy for fabrication of the different equipment is given by weight (excluding tyres).

(b) The energy for repairs and maintenance is considered to be approximately 1/3 of the total repair energy cost for the entire life of the equipment. The values utilised come from the ASEA tables given in Pimentel (1980).

(c) The useful working life corresponds to 82% of the total life of the equipment and the energy costs are converted to annual values based on these values.

The energy cost to manufacture steel from iron ore was reviewed by Worrell et al. (1997). The data for 1991 show that the energy cost in all countries except China, range from 20 to 30 GJMg″1. A later publication by Farla and Blok (2001) cite the World Energy Council for 1995 with a mean value of 22 GJMg″1. For the purposes of calculating the fossil energy input for steel for agricultural machinery in this section, as well as for buildings and equipment in cane mills/distilleries we have adopted the value of 30 MJ kg″1.

A full explanation of how the energy required to manufacture and maintain agricultural machinery are given in Appendix 13.1 and the total value is estimates as 1,785.6MJ ha"1 yr"1.

TRANSPORT COSTS

The fossil energy cost of transport of the cane from the field to the mill/distillery depends on the mean distance travelled to and from the mill by the transporting vehicles, the capacity (Mg cane) of the transport vehicles and the consumption of diesel fuel per km. The 1990 report by the National Audit Tribunal on the bioethanol program states that transport of cane a distance of more than 30 km from the mill is uneconomic (TCU, 1990). Managers answering a quick telephone survey of four mills in S˜ao Paulo and two in Rio de Janeiro States, said their mean radius of transport was between 14 and 20 km. If cane were planted uniformly around a mill to a radius of 30 km, the mean distance to fetch cane would be 22 km, so we have used this value. The cane transporters are predominantly a truck with a trailer (known in the business as a "Romeo and Julieta") and they transport between 26 and 30Mg cane (Macedo et al., 2003). Using the mean of 28 Mg/transporter, 1 ha of cane (76.6Mg) will require 2.74 loads. When loaded their diesel consumption is approximately 1.6 kmL"1 (Macedo et al., 2003), and we have assumed that when empty this is 3 kmL"1. So for a 44 km round trip at a mean of 2.3 kmL"1, the diesel consumption to fetch one ha of cane will be 44 × 2.74/2.33 = 51.7 L. There are 5 harvests every six years so the annual diesel consumption per ha will be 43.1 L ha"1 yr"1. At a energy value of 47.73 MJ L"1, mean fossil energy consumption for transporting cane to the mill becomes 2058 MJ ha"1 yr"1.

Transport of raw materials to plantation and mill from suppliers: The most important quantities of materials to be transported across the country to the plantations are lime and fertilisers. While the fertiliser supplier's association does not to provide data on individual quantities of N. P and K applied to each crop, they do give total tonnage of fertilisers applied to each crop. For 2006 their data were 3.13 millionMg of fertilisers added to 7.37Mha of sugarcane, a mean of 425 kg fertiliser applied per ha"1. For lime Macedo (1998) assumed that at each replanting (every 6 years) 6Mg of lime were applied, a mean of 367 kg lime ha"1 yr"1. The total (792 kg ha"1) is close to that of the 800 kg ha"1 value of Pimentel and Patzek (2007) and to account for the small quantities of inputs to the factories (lubricants, reagents) and pesticides, we use the value of 820 kg ha"1 which must be transported. The great majority of mills/distilleries are in the south-eastern region, predominantly in S˜ao Paulo. Fertilisers come from the factories that are situated close to the coast or are imported. In the southeast lime comes predominately from the State of Minas Gerais, but the northeast and other states have their reserves also. The main port of S˜ao Paulo State, Santos, is 410 km by road from one of the largest cane growing areas of the state, near Ribeir˜ao Preto. In almost all case, except for a few mills in the Central West region, transport will be predominately by road and less than 500 km distance. Assuming that all transport of raw material (820 kg ha"1) is in trucks hauling 35Mg for a distance of 500 km and fuel consumption of these vehicles is 2 kmL"1, an input of diesel per ha is of 5.8 L or 276.8MJ ha"1 yr"1.

Factory Inputs

Vast amounts of energy are utilised in factory processing, for pumping water, crushing the cane processing the juice, fermentation and distillation of the ethanol. Pimentel and Patzek (2007) estimate that it requires 2.546 Gcal per 1000 L of anhydrous ethanol for steam production for direct heating and to drive the electricity generators. This is equivalent to 16.9GJ ha"1 yr"1. However, all Brazilian mills/distilleries are powered by steam generated from burning the bagasse (crushed cane stems). In fact Macedo (1998) estimated that in most mills there is a surplus of bagasse-derived energy of between 8 and 14%. In some regions (especially S˜ao Paulo State) this surplus may be used to generate electricity which is exported to the local grid (in 2005, 350MW were exported to the grid by cane mills – Coelho, 2005), or the excess bagasse is transported to nearby industries for power generation, or sometimes used to produce fibreboard. Some energy for orange juice extraction plants near Ribeir˜ao Preto (SP), use excess bagasse for power generation. The use of bagasse for all factory inputs means that there is no extra fossil energy required to power the mills/distilleries.

The most important input of fossil energy for the factories is in their construction. The company Dedini S.A. based in Piracicaba, S˜ao Paulo State is now responsible for the construction of approximately 80% of the new sugarcane mills/ethanol distilleries in the whole of Brazil. At present (2006/2007) there are 357 mills/ distilleries in operation for a total harvested area of 6.72 Mha. This means the average mill has a harvested area of 18,800 ha with an annual production of 1.4 millionMg of cane. As the harvest period is almost universally 180 days, this means average mill throughput is approximately 8,000Mg of cane per day.

Engineers from Dedini S.A. provided us with construction details of a modern mill/distillery with throughput of 2 millionMg cane year"1. As these mills rarely run at full capacity, this size of mill approximately represents an average size mill in Brazil. Obviously a large new mill will be considerably more energy efficient that smaller and/or older mills, and we take this into account by calculating all energy inputs in the construction of this 2 million ton/day standard mill as if it were functioning only at one third capacity. In other words the total energy involved in its construction will be three times higher per ha, than if we considered that it was functioning at 100% of capacity.

Buildings are regarded as having a useful life of 50 years and a maintenance energy cost of 4% per yr (Macedo, 1997). As the greatest energy input is in cement the energy value of this material is used. As fuel costs are such a large part of manufacturing costs, most cement companies have aggressive energy conservation programs and according to the International Energy Program (IEA, 1999) new manufacturing plants have reduced energy use by between 25 and 40% compared to 10–15 years ago. The report by the IEA (1999) gives a value of 6.61GJ for the energy required to produced 1Mg of cement, and other recent reports (Young et al., 2002 and Worrell and Galitsky, 2004) give somewhat lower values of 4.35 and 6.1 GJMg"1. We use the former higher value of 6.61 GJMg"1. So allowing for a 4% annual maintenance cost the total embodied energy for all buildings over a 50 year period is (1,600 + (0.04×50×1,600)) ´U 6.61GJ which becomes 31,730 MJ or 634.6MJ yr"1. As we assume that the mill serves to grind one third of 2 millionMg of cane per year, this becomes 0.952 MJMg cane milled, or 75.9MJ ha"1 yr"1. For the mild steel in the mill/distillery we have assumed that one third is in light equipment and thus subject to more wear and will have a lower useful life (10 yrs – Macedo, 1997). The remaining two thirds is considered to be in the structure of the mill, equipment and distillery, and thus will have a longer useful life (25 years).

The data for the standard mill provided by Dedini S.A. show that 410Mg of stainless steel was used, mainly in the distillery columns. Pimentel and Patzek (2007) give the embodied energy in stainless steel to be 2.39 times that in mild steel, so the value of 71.7MJ kg"1 was used for this material. The useful life of this material was assumed to be 25 years. The energy input for stainless steel in the factory was again calculated using the same procedure as for cement. Finally to account for on-site energy utilised in the construction, all values were increased by 12.5% as suggested by Hannon et al. (1978). The total energy requirement for factory buildings and equipment totalled 1898 MJ ha"1 yr"1.

ENERGY BALANCE

Within the fossil energy inputs in the agricultural operations, fertilisers, especially N fertiliser, are responsible for the largest contributions. The fact that in Brazil N fertiliser use is far lower than in just about any other cane growing area in the world, makes an important economy. If for example 150 kgNha"1 yr"1 (typical of most other countries) were used instead of the estimated 56.7 kg, the energy input would rise from 3060 to 8100 MJ ha"1 yr"1 increasing the total energy input in agricultural operations (including transport of cane and consumables) by 43%. Because of their complicated synthesis herbicides are extremely energy intensive and even though only a mean of 3.2 kg a.i. ha"1 yr"1 are applied, this is the second most important consumables input after fertilisers. Brazil is fortunate in that most of the country has over 1,000mm of rainfall a year, and the most productive cane-growing areas have over 1,300 mm of rain. For this reason only a very small area is irrigated so that there is effectively no energy input for irrigation.

The comparatively large input of fossil energy in the manufacture of agricultural machinery, and to a lesser extent, of human labour, show the importance of including these inputs, which is not universal practice in computing such balances (e.g. Sheehan et al., 1998; Shapouri et al., 2002)

As all factory energy is supplied by bagasse, the main fossil energy input (estimated to be <"1,900 MJ ha"1 yr"1) is in the infrastructure of the construction and maintenance of structure and equipment of the factory. All factories are built near abundant water supplies (usually rivers) and pumping comes from electricity generated from bagasse, and thus involves minimal fossil energy inputs.

The total energy balance is the Total Energy Yield (TEY) of the biofuel divided by the Fossil Energy Invested (FEI). For today's production levels and practice we calculate this to be approximately 8.8, which is close to the value of 9.2 calculated for ethanol production S˜ao Paulo by Macedo (1998), and of 8.3 by Macedo et al. (2003). The main differences between these studies are that (a) we included the manual labour energy input, which was not included in the studies by Macedo and his colleagues, and (b) we used a much more recent estimate for the energy embodied in steel (30 MJ kg"1 –Worrell et al., 1997), rather than that cited by Macedo et al. (2003) of 38–63 MJ kg"1 which are estimates that date from the 1970s. When the energy balance (TEY/FEI) is high, differences of 1 or 2 units in the this ratio make only small differences in the proportion of energy saved.

Pimentel and Patzek (2007) estimated the input of fossil energy to produce Brazilian bioethanol was 13,286 MJm"3 (3,177 Mcalm"3) and a total energy yield of 21,454 MJm"3 (5,130 Mcalm"3). The resulting energy balance of 1.66 is in wide disparity of those calculated by Macedo (1998) and Macedo et al. (2003) and by us in this present study. There is a huge disparity in the estimates of the energy attributed to transport of consumables (fertilisers and chemicals for the factory) and of hauling cane from the field to the mill, the estimates of Pimentel and Patzek (2007) being respectively 10 and 33 times higher than ours. The consumption of diesel oil estimated by these authors seems totally unrealistic in that if a truck and trailer can carry 34Mg of cane for a 16 km round trip (their value) then the consumption of diesel at 47.7MJ L"1 would be 21.6 L km"1.

The energy balance computed by de Oliveira et al. (2005) of 3.7 is also considerably lower than that computed in this present study or Macedo (1998) and Macedo et al. (2003), and again the large difference comes in the utilisation of diesel fuel in the field operations and cane transport. These authors cite a report from the University of Campinas (Unicamp, Campinas, S˜ao Paulo State) for a value of 600 L of diesel fuel consumed per ha per year compared to a total of 71.2 L ha"1 yr"1 (43.1 L for cane transport, 22.3 L for field operations and 5.8 L for transport of consumables to the plantation/mill) in our study. Substituting our value for diesel consumption in the energy balance of de Oliveira et al. (2005) becomes 7.0.

GREENHOUSE GAS EMISSIONS

For the ethanol production, fossil fuel is used directly and indirectly for construction of the infrastructure of machinery and consumables together with other chemical and biological processes which are used in sugarcane production. The use of these fossil fuels results in the generation of greenhouse gases (GHGs). The energy data and amounts of material for factories, consumables, machinery, fuels and labour involved in the ethanol life cycle were used to estimate GHGs emissions based on emission factors for each component.

Inputs for agricultural operations are calculated from energy in labour, herbicides, insecticides and seeds which come from many different sources and they were assumed to be best represented by crude oil. From IPCC (1996) 1 GJ of crude oil emits 73.3 kg CO_2, 0.003 kg CH4 and 0.0006 kg N_2O. Estimates from machinery were based on the energy contained in the steel, which was assumed to come from steel factories fuelled by coking coal (1 GJ is equivalent to 94.6 kg CO2, 0.001 kg CH_4 and 0.0015 kg N_2O). Diesel oil was the energy source for transport of consumables and cane to the factory, fuel for machines and irrigation, which meant each GJ employed emitted 74.1 kg CO_2, 0.003 kg CH4 and 0.0006 kg N_2O (IPCC, 2006). Fertilisers and lime complete the components of sugarcane production. In the absence of information regarding the type of lime used in sugarcane areas (proportions of calcitic and dolomitic) emissions of CO_2 from lime addition were estimated by the amount of lime multiplied by the emission factor of 0.75, proposed by the IPCC (2006), tier 1. For fertiliser emissions, urea, triple superphosphate and potassium chloride were considered to best represent the NPK formulation used in sugarcane areas. The contribution of each source was estimated by the emission factors proposed by Kongshaug (1998). Assuming the average for this technology in Europe, the production of 1 kg of urea, 1 kg of triple superphosphate and 1 kg of potassium chloride represent 0.61, 0.17 and 0.34 kg CO_2 emitted to the atmosphere, respectively.

After N fertiliser placement (56.7 kgNha"1) and vinasse application (23 kgNha"1), it was assumed no NH+ was substrate for nitrification and denitrification processes for N_2O emissions. No significant CH4 production was considered to occur in the sugarcane areas during cropping phase (Macedo, 1998). The harvested area after burning was assumed to be 80% of the whole cropped area. In this case, fractions of 0.005 of total C (5.25Mg ha"1) and 0.007 of total N (30 kgNha"1) in burned trash were considered to evolve as CH4 and N_2O, respectively (IPCC, 2006). For the remaining 20% in unburned areas, the 30 kgNha"1were considered to be in harvest residues left to decompose in the field which meant a fraction of 0.0125 of this N was emitted as N_2O.

For factory construction and function the emissions coming from cement, steel and chemicals were accounted for, all based on emission factors from the IPCC guidelines (IPCC, 1996). For cement a factor of 0.95 was applied to calculate clinker content from the total cement used. According to Tier 1 this carries an emission factor of 0.507, with a 2% correction for cement kiln dust, and this was used to calculate the CO_2 emission. In the case of structural and mild steel, emissions of CO_2 were calculated on the basis of the global average emission factor for iron and steel production (1.06 kg CO2 kg"1 steel produced). For stainless steel, the emission factor for ferrochromium of 1.6 kg CO_2 kg"1 steel produced was used (IPCC, 2006). Energy in the production agro–chemicals was considered to be from crude oil for which the emission factors for CO_2, N_2O and CH_4 were mentioned above.

To explain the impact of ethanol from sugarcane produced under Brazilian conditions the agricultural activities were broken down into three different phases: planting, crop management and harvesting, the latter including transportation of cane to mill. The factory phase was also included to close the cycle. Emissions of CO_2 predominated at planting and were explained by the fossil fuel energy used in consumables, machinery and transportation of consumables. During plant development N_2O production was derived from fertiliser and vinasse N and nitrification/ denitrification gained importance and represented a large share (85%) of the emissions expressed as equivalents of CO_2. Again, the trace gases CH_4 and N_2O represented most of the emissions at harvest, the former was emitted mostly from burning trash at harvest and the latter, partially from burning, but also from decomposition of N in residues in unburned areas (20% of all Brazilian cane). The most important greenhouse gas (GG) emissions are incurred during pre-harvest burning, and amount to 82 kg and 588 kg ha"1 yr"1 of CO_2 equivalents, as N_2O and CH_4, respectively, 34% of all GG emissions.

The conversion from manual harvesting of burned cane to machine harvesting of green cane would eliminate these emissions as well as approximately 70% of the yearly manual labour input (0.7 × 1004 MJ ha"1 or 52 kg CO_2 equivalents ha"1). However, the decomposing trash emits 183 kg CO_2 equivalents ha"1 yr"1 as N2O from the 30 kgN left in the cane trash. Furthermore, the harvester (70Mg of cane harvested per h, machine weight 19 Mg) consumes 40 L of diesel per h (data from Sr. Aureo Tasch, John Deere S.A., Catal˜ao, Goi´as) giving a fossil energy input of 2089 MJ ha"1 yr"1 (155 kg CO2 equivalents ha"1 yr"1). Embodied energy in the machine (effectively 100% steel, 5.5 kg ha"1 yr"1) is equivalent to 54.2MJ (5.1 kg CO_2 equivalents ha"1 yr"1). In summary, under manual harvesting annual GG emissions amount to 722 kg CO_2 equivalents ha"1 and this falls to 343 kg CO_2 equivalents ha"1 if the cane is harvested green with machine harvesting. It is also reported that full ground cover with trash during the year reduces the requirement of herbicide by at least 50% (Antˆonio Gondim, Usina Cruangi, Timba´uba, Pernambuco; pers. comm.) equivalent to 60 kg CO_2 equivalents ha"1 yr"1. Emissions derived from the factory infrastructure and chemicals for ethanol production from milled cane accounted for less than 5% of the emissions calculated for the whole cycle.

Summing up: all emissions in terms of CO2 equivalents amount to approximately 2.36Mg CO_2 ha"1 yr"1, close to one fourth of the total emissions avoided whether burning ethanol as a fuel (9.58Mg CO_2 ha"1 yr"1), assuming 100% is converted to CO_2.

LOCAL AND REGIONAL IMPACTS

Atmospheric Pollution

Harvesting of cane occurs mainly in the dry season. The pre-harvest burning of the cane facilitates the manual harvest and diminishes the risks of injury to workers from snakes and poisonous spiders. The burning releases inhalable particles with a great number of components, most of all are carbon-rich alumino-silicate based, causing a typical overcast fog-like atmosphere, widespread in the cane districts in this season. However, the occurrence, composition and persistence of the smoke is highly dependent on specific weather conditions and this coincides with admission to hospitals of children and elderly people with respiratory problems (Godoi et al., 2004; Arbex et al., 2007). Some studies state that this effect is regarded as similar to what could be observed in urban areas exposed to industrial and automotive pollution, but also acknowledge that the ethanol addition to gasoline has contributed to decreasing air pollution, at least in the last twenty years, in the urban centres (Canc¸ado et al., 2006). As mentioned before biggest cane producer, S˜ao Paulo State has passed a law to regulate cane burning since 2003. The law defines what areas are able to use mechanical harvesting due to field slope, and sets a timetable. All the plantations under 12% of inclination should be totally mechanised by 2022. Cane burning in the other areas should be eliminated by 2032, when all areas ought to be harvested without burning.

Water Pollution

In the early years most distillery waste was disposed of without treatment into local rivers. As the waste usually contains approximately 1% soluble C and high levels of K, S and N and some P, the results were disastrous. Many rivers became eutrophic, there was massive death of fish and all other aquatic organisms, and the stench of this disposed waste could be scented many kilometres before arriving at a distillery. At first the factory owners were loath to return these wastes to the field as there was often an initial wilting of the cane leaves and signs of damage to the plants (Boddey, 1993). However, it was shown in many experiments that the plants soon recovered and benefited from the extra nutrients, and pumping the waste out onto the fields diluted with other wash water from the mills (which also had significant BOD), was a cheaper source of nutrients than synthetic fertilisers. Today almost all vinhac¸a is disposed of by pumping onto the fields, and where State and/or Municipal governments have effective environmental protection agencies, significant water pollution is a thing of the past.

Soil Erosion

Several authors have stated that soil erosion in sugarcane fields is major problem. Pimentel and Patzek (2007) write "Sugarcane production causes more intense soil erosion than any crop produced in Brazil because the total sugarcane biomass is harvested and processed in ethanol production". They cite the paper of Sparovek and Schnug (2001) who give a value for soil loss of 31Mgha"1 yr"1. This estimate was derived from the use of the Universal Soil Loss Equation on just one site near Piracicaba (S˜ao Paulo) with a slope of 5–15% and not based on actual soil loss measurements. Fuller details are given of this study in the paper of Bacchi et al. (2000). Actual measurements at this site using the 137Cs radioactive isotope technique (Ritchie and McHenry, 1990, 1995) yielded a mean value for soil loss of 23Mgha"1 yr"1. This technique yields mean annual values from approximately 1962 to the time of sampling. At the start of the 1960s there was a large increase in 137Cs deposition due to many very large nuclear explosions from H-bomb tests by the USSR and USA.

Only one other study using this technique seems to have been published (Correchel, 2003), and this author reported a mean annual soil loss of 10.8Mg ha"1 yr"1 on a 4% slope on an Oxisol. Lombardi-Neto et al. (1982) conducted a study where actual soil loss was measured from plots in a cane field on a 12.8% slope on a "latosolo roxo" (Typic Haplorthox – US Soil Taxonomy classification) for the plant crop and 2 ratoons. In the year when the soil was deep ploughed for planting losses were high (49Mgha"1) but for the subsequent two ratoon crop years losses were minimal (0.20 and 0.01Mgha"1) giving a mean loss of 16.4Mg ha"1 yr"1. Other estimates using the Universal Soil Loss Equation give values between 3.3 and 7.3Mg yr"1 of soil loss on slopes from between 3 and 8%.(de Souza et al., 2005).

In summary, as the crop is only renovated every 6 years, mean annual losses are generally lower than for other crops (e.g. soybean, maize) grown with conventional tillage, but a global figure for Brazil is not available, and where cane is grown on flat land losses will be much lower than in the above studies. Already 30% of the area of sugar cane in S˜ao Paulo State is being harvested without burning (green-cane harvesting) and the preservation of the cane trash on the soil surface in these areas will undoubtedly radically reduce erosion losses. While erosion losses are at present are moderate to severe, the increasing use of direct planting (zero tillage) of cane and green-cane harvesting in the next decade or so, these losses should fall to acceptable levels.

Replacement of Food Crops and Invasion of Reserves of Biodiversity Two major criticisms have been levied against the Brazilian ethanol program that need to be answered:

One can be summarised briefly as "the expansion of the ethanol program will increase the destruction of the Amazon forest". The government has declared recently that cane factories will not be licensed in reserves of biological diversity such as Amaz^onia and the Pantanal, and the data indicate that at present such areas are not threatened.

The other criticism is that sugar cane will replace food crops, especially those grown by subsistence farmers, leading to food shortages for the poorer sections of Brazilian society. The expansion of sugar cane is occurring principally onto areas purchased from large landowners/ranchers who have extensive areas of degraded pastures. It is estimated that in the Atlantic coastal region (which includes the States of S˜ao Paulo and Paran´a where expansion of cane is most rapid) there are perhaps 20 Mha of degraded pastures (Boddey et al., 2003) and in the Cerrado (central savanna) as much as 40 Mha of similar under-utilised, but not infertile land (Sano et al., 2001; Boddey et al., 2004). Brazil has no shortage of land for crop production. The resource-poor and landless require land reform, which is now occurring at an increased pace, but more importantly they need resources and skills to invest in the land and markets for their products and these facilities are only slowly becoming available. A vigorous rural economy fuelled by the intensive production of soybean, maize and sugarcane is more likely to provide employment for the poor, than the vast abandoned tracts of badly managed ranches which dominate the Cerrado region.

Labour Conditions

In this publication we have restricted comments on the social impact of the Bioethanol program to labour conditions. Approximately 80% of the area of Brazilian sugar cane is still harvested by hand. It is one of the most arduous occupations that exists in any industry. In cutting the burned cane the workers are exposed to the charred residues and they immediately become covered in ash and soot. To protect themselves from the rough cane stalks they must be fully attired with heavy protective clothing and boots and leggings to avoid injury with the sharp heavy cutlasses, while often working in temperatures which can reach 40°C. Even to cut the minimum requirement of one "tarefa" (<"6Mg of cane) requires an immense amount of energy and most workers manage to cut considerably more than this each day. The only reason that workers will accept such employment is because compared to most other work in the rural areas it is relatively well paid. Most workers manage to earn between R$ 600 and R$ 900 (US$ 300–US$ 450 – Globo Rural, 26 August 2007) per month which compares well to the national minimum wage of R$ 380, which is not often attained by the majority of workers in other rural occupations. According to the last national census conducted by the Brazilian Institute of Geography and Statistics (IBGE, 2004), sugar cane involves the direct employment of more than 251,000 permanent and 242,000 temporary employees, a total of about 493,000 agrarian workforce. Rural labour is almost always the only possible livelihood for the unqualified worker. The number of Brazilian agrarian workers officially registered in the Ministry of Work and Employment is 32.3%, which means full access to public health, working rights and a secured retirement. However permanent and temporary rural employees engaged in the sugar cane industry are 64.9% and 39.7%, respectively. That means even the temporary workers possess guarantees and a formal employment on a wider scale than the remainder of the national agricultural work force (Balsadi, 2007). However, there have been some cases of terrible working conditions and low wages at some mills, with isolated cases where the workers are paid only sporadically or only at the end of the harvest, such that they owe so much money to the "company store" that they are effectively slaves. There is a major campaign of the Federal authorities to eliminate such practices and they levy large fines on these employers, but in such a vast country these abuses still occur.

The cost of harvesting burned cane using manual labour is estimated to be approximately R$ 1000.oo ha"1, considerably more than machine harvesting which is estimated at R$ 700.oo ha"1, including the costs of the equipment and maintenance etc. For this reason those companies that have the financial resources are investing in harvesters (cost approximately R$ 800,000.oo) such that it is estimated that 450 machines will be sold in 2007 (Globo Rural, 26 August 2007). As one machine can harvest approximately 60Mg h"1 (18.7 ha per day – Anon 2005) These machines operate 24 h in three 8 h shifts and the harvest last 180 days. Assuming a operational efficiency of 80%, this year approximately 1.2Mha will be converted from manual harvesting of burned cane to machine harvesting of green cane. This will not only reduce operational costs, but also promote soil organic matter accumulation (Boddey et al., 2006), reduction in atmospheric pollution and methane emissions from burning, and improve soil fertility. The downside is that for each machine introduced approximately 80 jobs will be lost, increasing rural unemployment by 72,000. While manual cane harvesting may be a terrible task, it is far preferably to being unemployed.

CONCLUSIONS

It is unfortunate that the private companies, sugarcane co-operatives and even research institutions engaged in producing bioethanol, or studying the bioethanol program, rarely publish information in English and much of that in Portuguese is in unpublished reports and other grey literature. For this reason there is a lack of easily-available information on the history, growth and environmental and social impact of this large biofuel program. We have attempted to search out as much information as possible on this subject with special consideration for an international readership.

After relating the history of the program and the present situation today with regard to scale and agronomic and industrial practices, we have attempted to evaluate principally its environmental impact, both local and global, with regard to the ratio of fossil energy inputs to total energy yield of the fuel produced (energy balance), to its role in the mitigation of greenhouse gas emissions and other impacts such at water and atmospheric pollution. For the energy balance and greenhouse gas emissions we have tried to find the latest data not only on field and factory practice but also on the fossil fuel inputs and greenhouse gas emissions.

In agreement with other earlier reports from S˜ao Paulo State (Macedo, 1998; Macedo et al., 2003) we conclude that the energy balance is approximately 9:1; i.e. one unit of fossil energy invested produces 9 units of total energy as bioethanol. As this subject is controversial, we have tried to be as transparent as possible by clearly citing the sources used for our information. The greenhouse gas balance indicates that using bioethanol produced from sugarcane with present practice will result in a 79% abatement of greenhouse gas emissions compared to "pure" gasoline. By "pure" we refer to gasoline as used in the USA and other countries with no ethanol addition, but with MTBE. If a comparison with Brazilian "gasohol" (actually gasoline with a 22–24% ethanol addition) this greenhouse gas abatement becomes about 73%.

With regard to accusations that the expansion of the area of sugarcane in Brazil will induce destruction of the Amazon rain forest or other reserves of biodiversity, the evidence contradicts this. Only 7–8 Mha of land are used as present for sugarcane, compared to over 350 Mha of Amazon forest, and the new areas for cane cultivation are being established in the south-east and central-west regions of Brazil principally in areas of *Brachiaria* and similar pastures used for extensive cattle ranching.

Working conditions for the manually harvesting of cane are extremely severe, almost inhuman, but for the rural poor of many regions of Brazil such work is comparatively well paid and rates of formal employment (with health security and pension provision) are much higher than is general in the remainder of the rural sector.

In the future is seems inevitable that the industry and area planted to cane will grow. Also increased mechanisation is inevitable and far fewer manual workers will be required, which will have negative effects on employment but positive effects on salaries and working conditions within the industry. Increased mechanisation and the consequent abandonment of pre-harvest burning and the introduction of no-till planting will further increase the energy balance, and reduce greenhouse gas emissions as well as other forms of atmospheric pollution and soil erosion. If State and Federal environmental protection agencies become increasing effective and employment laws are enforced, the Brazilian bioethanol program will be of great economic, and environmental benefit to the country, and could play a small but significant global role in the mitigation of greenhouse gas emissions from motor vehicles and reducing the consumption of petroleum.

According to a recent report in the Newspaper Folha de S˜ao Paulo on average a Flex fuel car will need 10 L of hydrous ethanol (95% alcohol) to cover the same distance as it would cover with just 7.2 L of Brazilian gasohol.

11

Ethanol Production: Energy and Economic Issues

This analysis employs the most recent scientific data for the U.S. and for Brazil sugarcane production and the fermentation/distillation. These two countries were selected because they are the two largest countries in the world producing ethanol. All current fossil energy inputs used in the entire process of producing ethanol from sugarcane were included to determine the entire energy cost for ethanol production. Additional costs to consumers, including federal and state subsidies, plus costs of environmental pollution and/or degradation associated with the entire production system are discussed. The economic and the broad human food supply issues are evaluated. In addition, other studies are compared.

INTRODUCTION

The supply of "conventional" oil is projected to peak before 2010 and its decline thereafter cannot be compensated fully by other liquid fuels (Youngquist and Duncan, 2003). The United States, Brazil, and other nations critically need to develop liquid fuel replacements for oil in the near future. The present search for alternative liquid fuels has focused on the conversion of biomass into liquid fuels. Biomass is green plant material, like corn, soybeans, sugarcane, and trees. All biomass converts solar energy into plant material. However, the major difficulty in relying on the use of solar energy collected by plant biomass is that green plants on average, collect only about 0.1% of the solar energy that reaches the land each year (Pimentel et al., 2007). In addition, biomass production requires large land areas, suitable soil, nutrients, and freshwater for its production. Then, in the conversion of the biomass into liquid fuel, water, microorganisms, and energy are required.

Professor Mario Giampietro (personal communication, visiting Scholar, Arizona State University, 2007) reported that a person in a developed economy, like the U.S. and Brazil, requires a range of 10–100 W/m^2 of fossil energy in urban land uses. Without including any charges for the fossil energy inputs for sugarcane production in Brazil, the collection of solar energy in Brazil is equal to only 1.58W/m^2, but with the outcome of producing ethanol, then only 0.4 W/m^2 are captured. Therefore, sugarcane and corn and other biomass resources will not supply a developed economy sufficient energy (Patzek and Pimentel, 2005).

ENERGY INPUTS IN SUGARCANE PRODUCTION

The conversion of sugarcane and other food/feed crops into ethanol by fermentation is a well-known and established technology. In both the U.S. and Brazil, the energy costs for the production inputs and processing inputs were apportioned to each activity. Also, in both countries, the sugarcane bagasse that remains after crushing is burned to provide steam and electricity in the processing activity. The ethanol yield from a large production plant averages about 1 L of ethanol from 12 to 14 kg of fresh sugarcane.

UNITED STATES

The production of sugarcane in Louisiana, United States requires a significant energy and dollar investment for the 12 inputs, including labor, farm machinery, fertilizers, pesticides, and electricity. To produce an average sugarcane yield of 88,000 kg/ha requires the expenditure of about 13 million kcal of energy inputs (mostly oil and natural gas). This energy input is the equivalent of about 1,300 kg of oil equivalents expended per hectare of sugarcane. The full production costs total about $2,524/ha for the 88,000 kg/ha or approximately 3c/kg of sugarcane produced.

Brazil

The production of sugarcane in Brazil also requires a significant energy and dollar investments for the 12 inputs. Energy investment amounts to about 8.8 million kcal, to produce the average sugarcane yield of 77,000 kg/ha or slightly lower than in Louisiana. This is the equivalent of about 393 kg of oil equivalents expended per hectare of sugarcane produced in Brazil.

ENERGY INPUTS IN FERMENTATION/DISTILLATION

In order to be mixed with gasoline in the U.S. and Brazil, the 95% ethanol must be further processed to 99.5% pure ethanol. More water must be removed, requiring additional fossil energy inputs to achieve the 99.5% pure ethanol. Thus, a total of about 10 L of wastewater must be removed per liter of ethanol produced. This relatively large amount of sewage effluent has to be disposed of at an energy, economic, and environmental cost. To produce a liter of 99.5% ethanol from sugarcane in the U.S. uses less fossil energy than the energy delivered back as ethanol (input/output ratio 1 kcal: 1.48 kcal) and costs 57c per liter ($2.15 per gallon). The sugarcane feedstock requires more than 40% of the total energy input in processing. In this analysis the total economic cost, including the energy inputs for the fermentation/distillation process and the apportioned energy costs of the stainless steel tanks and other industrial materials, is $570 per 1000 L of ethanol produced. The production of 1,000 L of ethanol under Brazilian conditions uses less energy than in the U.S. With Brazilian ethanol production there is a net return of 2.28 kcal per 1 kcal of fossil energy invested. Yet, economic cost of production is 26c/liter compared to 57c/liter in the U.S.

Energy Yield

The largest energy inputs in sugarcane-ethanol production are for producing the sugarcane feedstock, plus the transport energy. The total energy input to produce a liter of ethanol in the U.S. is 3,458 kcal and in Brazil 2,242 kcal. A liter of ethanol has an energy value of 5,130 kcal. Thus, in the U.S. there is a positive energy return (1:1.48), while in Brazil the positive energy return is 1:2.28 or slightly larger than in the U.S.

Economic Costs

Not only does the U.S. ethanol production technology with sugarcane use nearly as much fossil fuel to produce as there is energy in the ethanol produced, but it also costs substantially more dollars to produce than its energy value is worth on the U.S. market. Clearly, without the more than $3 billion U.S. federal and state government yearly subsidies, U.S. corn or sugarcane ethanol production would be reduced or cease to exist, confirming the basic fact that ethanol production is uneconomical (National Center for Policy Analysis, 2002).

Federal and state subsidies for corn ethanol production total more than $7 per 25 kg (1 bushel), for example. The subsidies are mainly paid to large corporations, like Archer Daniels Midland, and others (McCain, 2003), while corn farmers, for instance, are receiving a maximum of only about 70c per 25 kg (1 bushel) for their corn (or about $100 per acre) in the subsidized corn ethanol production system. Senator McCain reports that direct subsidies for ethanol, plus the subsidies for corn grain, amount to 79c per liter of ethanol produced (McCain, 2003). Thus, the subsidy per liter of U.S. ethanol is 60 times greater than the current subsidy per liter of gasoline, based on the $5 billion per year subsidy for oil (Gara, 2006; NCGA, 2006; Koplow, 2006).

If the production cost of a liter of ethanol were added to the tax subsidy cost, then the total cost for a liter of sugarcane ethanol would be $1.11. Because of the relatively low energy content of ethanol, 1.6 L of ethanol have the energy equivalent of 1 L of gasoline. Thus, the cost of producing an amount of ethanol equivalent to a liter of gasoline is $1.78 ($6.71 per gallon of gasoline). This is more than the 53c per liter, the current cost of producing a liter of gasoline (USCB, 2004–2005).

The subsidies for ethanol produced from sugarcane in the U.S. would have similar effects as subsidies in corn ethanol. Unfortunately the costs to the American consumer are greater than the $8.4 billion/year expended to subsidize and produce the U.S. ethanol because diverting the required corn feedstock from livestock increases corn grain prices for livestock producers. The National Center for Policy Analysis (2002) estimate is that ethanol production is adding more than $1 billion per year to the cost of beef production for consumers (2002). Given that about 78% of the current corn grain harvest is fed to U.S. livestock (USDA, 2004), the doubling or tripling of ethanol production can be expected to increase corn grain prices further for beef production and for other livestock products, including milk and eggs, and ultimately increase costs to the consumer. Therefore, in addition to paying the $8.4 billion in taxes for ethanol and subsidies, consumers are expected to face significantly higher meat, milk, and egg prices, plus higher sugar prices if much sugarcane eventually goes into ethanol production in the market place.

Land Use in the U.S.

Currently 395,000 ha are devoted to U.S. sugarcane production and most of t' is is for sugar production (USDA, 2004). Relatively little or no ethanol is being produced from U.S. sugarcane. Currently, about 18.9 billion liters of ethanol (5 billion gallons) are being produced in the United States each year primarily from corn (DOE, 2005). In contrast, the total petroleum used in the U.S.was about 1,200 billion liters in 2004–2005 (USCB, 2004–2005). To produce the 18.9 billion liters of ethanol, about 5.0 million ha or 20% of U.S. corn land is used. Furthermore, the 18.9 billion liters of ethanol (energy equivalent to 12.5 billion liters of vehicle liquid fuel) provides only 1% of the petroleum utilized by U.S. each year. If corn-ethanol production were expanded to using 100% of U.S. corn production, this would provide only 7%of the petroleum needs!

ETHANOL PRODUCTION AND USE IN BRAZIL

In contrast, Brazil can fuel most of its automobiles and other vehicles with ethanol because Brazilians consume only 9% of the U.S. consumption in petroleum (BP, 2005). Since 1984 the portion of Brazilian sugarcane used for ethanol decreased from a peak of 70% to about 55% in 2000 (Schmitz et al., 2003). During that time the percentage of ethanol cars declined from 94% in 1984 to less than 1% in 1996 (Rosillo-Calle and Cortez, 1998). The difference included gasoline for the cars. Flex cars replaced the ethanol cars and as a result Brazil's oil consumption has increased 42% during the last decade (BP, 2001, 2005).

Proponents of ethanol point to the production of ethanol in Brazil but ignore the fact that the U.S. now produces more ethanol (18.9 billion liters ethanol per year) compared with Brazil that produces about 15.1 billion liters per year (Calibre, 2006). Brazil is fortunate to have the land and climate suitable for sugarcane. Sugarcane is a more efficient feedstock for ethanol production than corn grain (Patzek and Pimentel, 2005). However, because Brazilian energy balance is only slightly positive (1 kcal:2.28 kcal), the Brazilians need to heavily subsidize their ethanol industry as does the U.S.. In the 1980s and 1990s the Brazilian government sold ethanol to the public for 22c per liter, but it cost the government 33c per liter to produce (Pimentel, 2003). Because other priorities emerged in Brazil, the government has since abandoned directly subsidizing ethanol (Spirits Low, 1999; Coelho et al., 2002). Now the consumer is paying the subsidy directly at the pump (Pimentel, 2003). The total Brazilian subsidy is estimated to be about 50% for ethanol production (CIA, 2005). Earlier it was mentioned that it costs 26c to produce a liter of ethanol in Brazil that sells for 86c per liter (Calibre, 2006). Brazilian gasoline sells for nearly $1.23 per liter or about 43% higher than a liter of ethanol (R.M Boddey, Senior Scientist, Empresa Brasileira de Pesquisa Agropecuaria (Embrapa), Brasil, personal communication, 2007). Thus, higher gasoline prices help subsidize the cost of ethanol production in Brazil (CIA, 2005).

Environmental Impacts

Some of the economic and energy contributions of ethanol production both in the U.S. and Brazil are negated by the widespread environmental pollution problems associated with ethanol production using sugarcane. Many of the environmental impacts in Brazil associated with sugarcane production also occur in the U.S. sugarcane production. Sugarcane production causes more intense soil erosion than any crop produced in Brazil because the total sugarcane biomass is harvested and processed in ethanol production. This removal of most of the biomass leaves the soil unprotected and exposed to erosion from rainfall and wind energy. For example, soil erosion with sugarcane cultivation is reported to have the highest soil erosion rate in all Brazilian agriculture, averaging 31 t/ha/yr (Sparovek and Schung, 2001). The 31 t/ha soil loss is 30–60 times greater than sustainability of the soil in agriculture (Troeh et al., 2004; Pimentel, 2006).

In addition, sugarcane production uses larger quantities of herbicides and insecticides and nitrogen fertilizer than most other crops produced in Brazil and these chemicals spread to ground and surface water thereby causing significant water pollution (NAS, 2003).

Relatively large quantities of water are required to produce sugarcane. Because it takes 12 kg of sugarcane to produce 1 L of ethanol, about 7,000 L of water are needed to produce the required 12 kg of sugarcane per liter of ethanol. Although the Brazilian government has passed legislation to curtail the burning of sugarcane before harvest to reduce air pollution problems, most of the sugarcane in Brazil is still burned and this is resulting in respiratory problems in children and the elderly (Braunbeck et al., 1999; Cancado et al., 2006). The rules need to be enforced to help protect the people from this serious air pollution problem. Additional smoke is released during the removal of forests for sugarcane and other crop production. Between May 2000 and August 2005, Brazil lost more than 132,000 square km of forest, an area larger than Greece (Mongabay, 2006).

The harvesting of sugarcane by laborers is hard and dangerous work, cutting the sugarcane with large knives. As Broietti (2003) reported these are dangerous and miserable conditions under which to work.

All these factors confirm that the environmental and agricultural system in which Brazilian and U.S. sugarcane is being produced is experiencing major environmental problems. Further, it substantiates the conclusion that the sugarcane production system, and indeed the ethanol production system, are not environmentally sustainable now or for the future. Because sugarcane is the raw material for ethanol production in Brazil, it cannot be considered a renewable energy source, considering the production and processing aspects.

Another pollution problem concerns the large amounts of waste-water produced by each ethanol plant. As noted, for each liter of ethanol produced using sugarcane, about 10 L of wastewater are produced. This polluting wastewater has a biological oxygen demand (BOD) of 18,000–37,000 mg/liter depending of the type of plant (Kuby et al., 1984). The cost of processing this sewage in terms of energy (4 kWh/kg of BOD) must be included in the cost of producing ethanol.

Air Pollution

Reports confirm that ethanol use contributes to air pollution problems when burned in automobiles (Youngquist, 1997; Hodge, 2002, 2003, 2005; Niven, 2005). The use of fossil fuels, as well as the use of ethanol in cars, releases significant quantities of pollutants to the atmosphere. Furthermore, carbon dioxide emissions released from burning these fossil fuels contribute to global warming and are a serious concern (Schneider et al., 2002). Additional carbon dioxide is released during the fermentation process. Also, when the soil is tilled serious soil erosion takes place and soil organic matter is oxidized. When all the air pollutants associated with the entire ethanol production system are considered, the evidence confirms that ethanol production contributes to the already serious U.S. and Brazilian air pollution problem (Youngquist, 1997; Hodge, 2002, 2003, 2005; Pimentel and Patzek, 2005; Patzek and Pimentel, 2005).

Food Security

At present, world agricultural land supplies more than 99% of all world food (calories), while aquatic ecosystems supply less than 1% (FAO, 2002). Worldwide, during the last decade, per capita available cropland decreased 20% and irrigation land 12% (Brown, 1997). Furthermore, per capita grain production has been decreasing, in part due to increases in the world population (Worldwatch Institute, 2001). Worldwide, diverse cereal grains, including corn, make up 80% of the food of the human food supply (Pimentel and Pimentel, 1996).

The current food shortages throughout the world call attention to the importance of continuing U.S. and Brazilian exports of grains and other food crops for human nutrition. The expanding world population that now numbers 6.5 billion, further complicates and stresses the food security problem now and for the future (PRB, 2006). Almost a quarter million people are added each day to the world population, and each of these human beings requires adequate food. Today, the malnourished people in the world number more than 3.7 billion (WHO, 2006). This is the largest number of malnourished people and proportion ever reported in history. Malnourished people are highly susceptible to various serious diseases and this is reflected in the rapid rise in the number of seriously infected people in the world, with diseases like tuberculosis, malaria, and AIDS, as reported by the World Health Organization (Kim, 2002; Pimentel et al., 2006).

FOOD VERSUS THE FUEL ISSUE

Using sugarcane, a human food resource, for ethanol production, raises ethical and moral issues (Wald, 2006). Expanding ethanol production entails diverting valuable cropland from the production of food crops needed to nourish people. The energetic and environmental aspects, as well as the moral and ethical issues also deserve serious consideration. In spite of oil and natural gas shortages now facing the U.S., ethanol production is forcing the U.S. to import more oil and natural gas to produce ethanol and other biofuels (Pimentel and Patzek, 2005).

The expansion of ethanol production in the U.S. and Brazil is having negative impacts on food production and food exports (Chang, 2006), and is likely to have further negative impacts on food production and the environment. Furthermore, increasing oil and natural gas imports in the U.S. and other countries drives up the price of oil and gas. This is especially critical for the poor in developing countries of the world. Even now this is documented by the fact that worldwide per capita fertilizer use has been declining for the last decade because of the increased costs for the poor farmers of the world (Worldwatch Institute, 2001).

SUMMARY

For a thorough and up-to-date evaluation of all the fossil energy costs of ethanol production from sugarcane in both the U.S. and Brazil, every energy input in the biomass production and ultimate conversion process must be included. In this study, more than 12 energy inputs in average U.S. and Brazilian sugarcane production are evaluated. Then in the fermentation/distillation operation, 9 more fossil fuel inputs are identified and included.

Some energy and economic credits are given for the bagasse to reduce the energy inputs required for steam and electricity.

Based on all the fossil energy inputs in U.S. sugarcane conversion process, a total of 1.48 kcal of ethanol is produced per 1 kcal of fossil energy expended. In Brazil, a total of 2.28 kcal of ethanol is produced per 1 kcal of fossil energy expended. Some pro-ethanol investigators have overlooked various energy inputs in U.S. and Brazilian sugarcane production, including farm labor, farm machinery, processing machinery, and others. In other studies, unrealistic low energy costs were attributed to such energy inputs, as nitrogen fertilizer, insecticides, and herbicides (Corn-Ethanol, 2007). Both the U.S. and Brazil heavily subsidize ethanol production. The data suggest that billions of dollars are invested in subsidies and this significantly increases the costs to the consumers.

The environmental costs associated with producing ethanol in the U.S. and Brazil are significant but have been generally overlooked. The negative environmental impacts on the availability of cropland and freshwater, as well as on air pollution and public health, have yet to be carefully assessed. These environmental costs in terms of energy and economics should be calculated and included in future ethanol analyses so that sound assessments can be made. In addition, the production of ethanol in the U.S. and Brazil further confirms that the mission of converting biomass into ethanol will not replace oil. This mission is impossible.

General concern has been expressed about taking food crops to produce ethanol for burning in automobiles instead of using these crops as food for the many malnourished people in the world. The World Health Organization reports that more than 3.7 billion humans are currently malnourished in the world – the largest number of malnourished ever in history.

12

Ethanol Production Using Corn, Switchgrass and Wood; Biodiesel Production Using Soybean

In this analysis, the most recent scientific data for corn, switchgrass, and wood, for fermentation/distillation were used. All current fossil energy inputs used in corn production and for the fermentation/distillation were included to determine the entire energy cost of ethanol production. Additional costs to consumers include federal and state subsidies, plus costs associated with environmental pollution and/or degradation that occur during the entire production process. In addition, an investigation was made concerning the conversion of soybeans into biodiesel fuel.

INTRODUCTION

Green plants, such as corn, soybeans, switchgrass and trees, and all other kinds of biomass, convert solar energy into plant material but require suitable soil, nutrients, and freshwater. In the conversion of the biomass into liquid fuel, water, microorganisms, and more energy are required. Andrew Ferguson (2006, personal communication, Optimum Population Trust, Manchester, UK) makes an astute observation that the proportion of sun's energy that is converted into useful ethanol, even using very positive energy data, only amounts to 5 parts per 10,000, or 0.05% of the solar energy.

Some recent papers are claiming returns on ethanol production from corn of anywhere from 1.25 kcal to 1.67 kcal per kcal invested (Shapouri et al., 2004; Farrell et al., 2006; Hill et al., 2006). These excessively high returns are achieved by either omitting several energy inputs, reducing other energy inputs, or giving credits that are too optimistic for the by-products.

ENERGY INPUTS IN CORN PRODUCTION

The conversion of corn into ethanol by fermentation in a large plant is about 1 liter of ethanol from 2.69 kg of corn grain (approximately 9.5 liters pure ethanol per bushel of corn. The production of corn in the United States requires a significant energy and dollar investment for the 14 inputs, including labor, farm machinery, fertilizers, irrigation, pesticides, and electricity. To produce an average corn yield of 9,400 kg/ha (149 bu/ac) of corn using up-to-date production technologies requires the expenditure of about 8.2 million kcal of energy inputs (mostly natural gas, coal, and oil). This is the equivalent of about <"930 liters of oil equivalents (<"25% of grain energy) expended per hectare of corn. The production costs total about $927/ha for the 9,400 kg/ha or approximately 10c/kg ($2.54/bushel) of corn produced. Full irrigation (when there is insufficient or no rainfall) requires about 100 cm/ha of water per growing season. Because only about 15% of U.S. corn production currently is irrigated (USDA, 1997a), only 8.1 cm per ha of irrigation was included for the growing season. On average irrigation water is pumped from a depth of 100m (USDA, 1997a). On this basis, the average energy input associated with irrigation is 320,000 kcal per hectare.

ENERGY INPUTS IN FERMENTATION/DISTILLATION

The average costs in terms of energy and dollars for a large (245 to 285 million liters/year), modern drygrind ethanol plant. In the fermentation/ distillation process, the corn is finely ground and approximately 15 liters of water are added per 2.69 kg of ground corn. Some of this water is recycled. After fermentation, to obtain a liter of 95% pure ethanol from the 8–12% ethanol beer and 92–88% water mixture, the 1 liter of ethanol must be extracted from approximately 11 liters of the ethanol/ water mixture. Although ethanol boils at about 78 degrees C, and water boils at 100 degrees C, the ethanol is not extracted from the water in the first distillation, which obtains 95% pure ethanol (Maiorella, 1985;

Wereko-Brobby and Hagan, 1996; S. Lamberson, personal communication, Cornell University, 2000). To be mixed with gasoline, the 95% ethanol must be further processed and more water removed, requiring additional fossil energy inputs to achieve 99.5% pure ethanol. Thus, a total of about 10 liters of wastewater must be removed per liter of ethanol produced, and this relatively large amount of sewage effluent has to be disposed of at an energy, economic, and environmental cost. To produce a liter of 99.5% ethanol uses 46% more fossil energy than the energy produced as ethanol and costs 45c per liter ($1.71 per gallon). The corn feedstock requires about 32% of the total energy input. In this analysis, the total cost, including the energy inputs for the fermentation/distillation process and the apportioned energy costs of the stainless steel tanks and other industrial materials, is $454.23 per 1000 liters of ethanol produced.

Net Energy Yield

The largest energy inputs in corn-ethanol production are for producing the corn feedstock, plus the steam energy, and electricity used in the fermentation/distillation process. The total energy input to produce a liter of ethanol is 7,474 kcal. However, a liter of ethanol has an energy value of only 5,130 kcal. Based on a net energy loss of 2,344 kcal of ethanol produced, 46% more fossil energy is expended than is produced as ethanol.

Economic Costs

Current ethanol production technology uses more fossil fuel and costs substantially more to produce in dollars than its energy value is worth on the market. Clearly, without the more than $3 billion federal and state government yearly subsidies, U.S. ethanol production would be reduced or cease, confirming the basic fact that ethanol production is uneconomical (National Center for Policy Analysis, 2002). Federal and state subsidies for ethanol production that total more than $6 billion/ year for ethanol are mainly paid to large corporations (Koplow, 2006), while corn farmers are receiving a minimum profit per bushel for their corn (Pimentel and Patzek, 2005). Senator McCain reports that direct subsidies for ethanol, plus the subsidies for corn grain, amount to 79c per liter (McCain, 2003).

If the production cost of a liter of ethanol were added to the tax subsidy cost, then the total cost for a liter of ethanol would be $2.47. The mean wholesale price of ethanol was almost $1.00 per liter without subsidies. Because of the relatively low energy content of ethanol, 1.6 liters of ethanol have the energy equivalent of 1 liter of gasoline. Thus, the cost of producing an equivalent amount of ethanol equal a liter of gasoline is $3.00 ($11.34 per gallon of gasoline). This is more than 53c per liter, the current cost of producing a liter of gasoline. The subsidy per liter of ethanol is 60 times greater than the subsidy per liter of gasoline! This is the reason why ethanol is so attractive to large corporations.

Cornland Use

Currently, about 18 billion liters of ethanol (5 billion gallons) are being produced in the United States each year (Kansas Ethanol, 2006). The total amount of petroleum fuels used in the U.S. was about 2,500 billion liters (USCB, 2004–2005). Therefore, 18 billion liters of ethanol (energy equivalent to 11.2 billion liters of petroleum fuel) provides only 1% of the petroleum utilized last year. To produce this 18 billion liters of ethanol, about 5.7 million ha or 20% of U.S. corn land is used. Expanding corn-ethanol production to 100% of U.S. corn production would provide just 7% of the petroleum needs of the U.S.

By Products

The energy and dollar costs of producing ethanol can be offset partially by byproducts, like the dry distillers grains (DDG) made from dry-milling of corn. From about 10 kg of corn feedstock, about 3.3 kg of DDG with 27% protein content can be harvested (Stanton, 1999). This DDG is suitable for feeding cattle that are ruminants, but has only limited value for feeding hogs and chickens. In practice, this DDG is generally used as a substitute for soybean feed that contains 49% protein (Stanton, 1999). However, soybean production for livestock feed is more energy efficient than corn production, because little or no nitrogen fertilizer is needed for the production of this legume (Pimentel et al., 2002). In practice, only 2.1 kg of soybean protein provides the equivalent nutrient value of 3.3 kg of DDG. Thus, the credit fossil energy per liter of ethanol produced is about 445 kcal (Pimentel et al., 2002).

Factoring this credit for a non-fuel source in the production of ethanol reduces the negative energy balance for ethanol production from 46% to 39%. The high energy credits for DDG given by some are unrealistic because the production of livestock feed from ethanol is uneconomical given the high costs of fossil energy, plus the costs of soil depletion to the farmer (Patzek, 2004). The resulting overall energy output/input comparison remains negative even with the credits for the DDG by-product.

Environmental Impacts

Some of the economic and energy contributions of the by-products are negated by the widespread environmental pollution problems associated with ethanol production. First, U.S. corn production causes more soil erosion that any other U.S. crop (Pimentel et al., 1995; NAS, 2003). In addition, corn production uses more herbicides and insecticides and nitrogen fertilizer than any other crop produced in the U.S., and these chemicals invade ground and surface water, thereby causing more water pollution than any other crop (NAS, 2003).

As mentioned, the production of 1 liter of ethanol requires 1,700 liters of freshwater both for corn production and for the fermentation/distillation processing of ethanol (Pimentel and Patzek, 2005). In some Western irrigated corn acreage, like some regions of Arizona, ground water is being pumped 10-times faster than the natural recharge of the aquifers (Pimentel et al., 2004).

All these factors confirm that the environmental and agricultural system in which U.S. corn is being produced is experiencing major degradation. Further, it substantiates the conclusion that the U.S. corn production system, and indeed the entire ethanol production system, is not environmentally sustainable now or for the future, unless major changes are made in the cultivation of this major food/feed crop. Because corn is raw material for ethanol production, it cannot be considered a renewable energy source.

Furthermore, pollution problems associated with the production of ethanol at the chemical plant sites are emerging. The EPA (2002) already has issued warnings to ethanol plants to reduce their air pollution emissions or be shut down. Another pollution problem concerns the large amounts of wastewater produced by each ethanol plant. As noted, for each liter of ethanol produced using corn, from 6–12 liters of wastewater are produced. This polluting wastewater has a biological oxygen demand (BOD) of 18,000 to 37,000 mg/liter depending of the type of plant (Kuby et al., 1984). The cost of processing this sewage in terms of energy (4 kWh/kg of BOD) was included in the cost of producing ethanol.

Basically the major problem with corn and all other biomass crops is that they collect on average only 0.1–0.2% of the solar energy per year. At a fairly typical gross yield of 3,000 liters of ethanol per hectare per year, the power density achieved is only 2.1 kW/ha. That is compared with the gross power density achieved via oil, after delivery for use, on the order of 2,000 kW/ha. (A.R.B. Ferguson, personal communication, Optimum Population Trust, November 6, 2005). If all the current 28 million hectares of corn production were to be devoted only to growing corn for ethanol, then this acreage would supply only 7% of U.S. petroleum needs (USDA, 2003).

FOOD SECURITY

At present, world agricultural land supplies more than 99% of all world food (calories), while aquatic ecosystems supply less than 1% (FAO, 2006). Worldwide, during the last decade, per capita available cropland decreased 20% and irrigation land 12% (Brown, 1997). Furthermore, per capita grain production has been decreasing, in part due to increases in the world population (FAO, 2006). Worldwide diverse cereal grains, including corn, make up 80% of the human food supply (Pimentel and Pimentel, 1996).

The expanding world population that now numbers 6.5 billion, further complicates and stresses the food security problem now and for the future (PRB, 2006). Almost a quarter million people are added each day to the world population, and each of these human beings requires adequate food. Today, the malnourished people in the world number about 3.7 billion (WHO, 2000). This is the largest number of malnourished people and proportion ever reported in history. Malnourished people are highly susceptible to various serious diseases. The World Health Organization reports a rapid rise in the number of people in the world who are infected with diseases like tuberculosis, malaria, and AIDS (Kim, 2002; Pimentel et al., 2006).

FOOD VERSUS THE FUEL ISSUE

Using corn, a basic human food resource, for ethanol production, raises ethical and moral issues (Pimentel and Patzek, 2005). Expanding ethanol production entails diverting valuable cropland from the production of corn needed to nourish people. The energetic and environmental aspects, as well as the moral and ethical issues also deserve serious consideration. With oil and natural gas shortages now facing the United States, ethanol production is forcing the U.S. to import more oil and natural gas to produce ethanol and other biofuels (Pimentel and Patzek, 2005). Furthermore, increasing oil and natural gas imports drives up the price of oil and gas; this is especially critical for the poor in developing countries of the world. The impact is documented by the fact that worldwide per capita fertilizer use has been declining for the last decade (FAO, 2006).

CELLULOSIC ETHANOL

Properties of Cellulose

The term "cellulosic ethanol" is imprecise. It is meant to suggest that certain components of wood and green plant materials (cellulose, pectins, and hemicelluloses) can be chemically separated (from mostly lignin in wood) and partially split into hexose and pentose monomers, which are then fermented to produce ethanol. Low energy industrial processes of ethanol production from biomass do *not* exist.

Cellulose is the principal structural component of cell walls in higher plants. It is the most abundant form of living terrestrial biomass (Pimentel, 2001). For hundreds of millions of years, cellulose has protected plants from elements and animals, and from chemical attacks by fungi and bacteria. Cotton is 98% pure cellulose; flax is 80%, and wood is 40–50% cellulose, with the remaining 50–60% made up from other complex polysaccharides (20–35% hemicelluloses and 15–35% lignin). The special properties of cellulose result from the association of the long, straight polymeric chains to form fibers called micro-fibrils, which are stronger than steel. The micro-fibrils then form larger fibers, which are laid down in a cris-cross pattern, and intermixed with gel-like polysaccharides, hemicelluloses and pectins, that function as biocement (Taiz and Zeiger, 1998). In some ways this structure resembles fiberglass and other composite materials, in which rigid crystalline fibers are used to reinforce a more flexible matrix.

The beta-glycosidic bonds are crucial in determining the structural properties of cellulose, and thus the strength of the cellulose fibers. Because of the beta-bonds, the chain assumes an extended rigid configuration, with each glucose residue turned 180 degrees from its neighbor (Taiz and Zeiger, 1998). Another consequence of alternating top/bottom glucose residues is that OH groups of adjacent chains allow very extensive hydrogen-bonding between chains. This extensive inter-chain hydrogen-bonding and rigid beta-configuration makes cellulose fibers very strong and able to resist strong sodium hydroxide and acid solutions. In summary, close to one billion years of plant evolution have made cellulose very stable and resistant to biochemical attacks. Cellulose can be quickly decomposed and hydrolyzed only by mechanical grinding or steam exploding and severe chemical attack by hot concentrated sulfuric acid or sodium hydroxide. Biochemical enzymatic attacks take a long time and have low efficiency. The process of separating cellulose fibers from the rest of woody biomass is well-known, fast, efficient, and very energy intensive. It is called the paper kraftprocess. The kraft process is used in production of paper pulp and involves the use of caustic sodium hydroxide and sodium sulfide to extract the lignin from the wood fiber in large pressure vessels called digesters. The process name is derived from German "kraft," meaning strong. Unfortunately, the best energy efficiency of this process is <"6200 kcal/kg of paper, more than the high heating value of pure ethanol. Therefore a much milder, enzymatic process must be used to obtain simple sugars from cellulose.

DISADVANTAGES OF CELLULOSIC ETHANOL

CONTAMINATION

For corn starch fuel ethanol, normal fermentation times in batch mode (there are no continuous reactors in operation) are 48 hours; up to 72 hours is acceptable. These estimates do not include downtime, cleaning, start up, etc. Over 72 hours the number of failures increases exponentially due to contamination with bacteria: acetogens and others. As described in the literature, typical enzyme processes for lignocellulosic alcohol take 5–7 days, i.e., about 120–170 hours. This spells big problems if lignocellulosic ethanol producers ever go outside the laboratory or pilot scale (sterile fermenters) to a conventional fermentation vessel, which can *not* be sterilized for 120–170 hours.

Biomass Availability

Natural *net* productivity of a mature ecosystem (an earth household, e.g., a forest or grassland) is low on human time scale. (Very slow carbon burial occurs on geological time scale.) What is produced by autotrophic plants, rock weathering and floods, is consumed by heterotrophs (bacteria, fungi, and animals that are continuously recycled as nutrients for the plants). Some bacteria and fungi, in return for the food from plant roots, capture nitrogen from the air and convert it to ammonia, thus providing natural fertilization. Therefore, "biowaste" is an engineering classification of plant (and animal) parts unused in an industrial process. This dated human concept is completely alien to natural ecosystems, which must recycle their matter completely in order to survive. Excessive "biowaste" removal robs ecosystems of vital nutrients and species, and degrades them irreversibly. As discussed in (Patzek and Pimentel, 2005), those ecosystems from which we remove biomass at high rate (crop fields, tree plantations) must be heavily subsidized with fossil energy and earth minerals.

Enzyme Yield vs. Rate

The rate of lignocellulose hydrolysis and fermentation can be increased by enough pre-treatment (such as ball milling to exceedingly fine dust, at enormous energy costs, or steam exploding with acid pre-treatment), but rates will slow down rather rapidly before high yields are obtained. The main problem is the number of binding sites available; the outside-in rate limitation phenomenon. It simply takes time to chew into the sturdy lignocellulosic particles. Of course, one could run the lignocellulose through the kraft-like process. This cannot be done, however, for lignocellulosic ethanol because energy losses would be severe. One can get rather good yields and rates if one performs energy-intensive and unaffordable pretreatment, or (relatively) high yields with modest pre-treatment if one waits long enough (ideally for weeks). Thus, despite claims to the contrary, a real industrialprocess for lignocellulosic ethanol does not exist, and may *never* have a sufficiently favorable energy balance.

Thermodynamics

Current energy efficiency of producing cellulosic ethanol is so low that all other investigated paths to liquid biofuels are better; see (Patzek and Pimentel, 2005).

SWITCHGRASS PRODUCTION OF ETHANOL

The average energy input per hectare for switchgrass production is only about 3.9 million kcal per year. With an exceptional average yield of 10 t/ha/yr, this suggests for each kcal invested as fossil energy the return is 11 kcal – an excellent return. This return is *impossible* to realize for more than one year in environments other than an ecologically-balanced prairie. Nonetheless, massive industrial *monocultures* of switchgrass are proposed and studied. If pelletized for use as a fuel in stoves, the return is reported to be about 1:14.6 kcal (Samson et al., 2004).

However, converting switchgrass into ethanol results in a negative energy return. The negative energy return is 68% or a slightly more negative energy return than corn ethanol production. The cost of producing a liter of ethanol using switchgrass was 93c. The two major energy inputs for switchgrass conversion into ethanol were steam and electricity production.

Wood Cellulose Conversion into Ethanol

The conversion of 5,000 kg of wood harvested from a sustainable forest into 1,000 liters of ethanol require an input of about 9.2 million kcal. Therefore, the wood cellulose system requires slightly more energy to produce the 1,000 liters of ethanol than when using switchgrass. About 81% more energy is required to produce a liter of ethanol using wood than the energy harvested as ethanol. This includes harvesting 25% of the wood lignin and filtering out the lignin from the residue after the ethanol has been removed by fermentation. The lignin residue was assumed to have 200% moisture and did not require further drying (thus, more energy) before being burned to produce steam.

The ethanol cost per liter for wood-produced ethanol is slightly higher than the ethanol produced using switchgrass, $1.01 versus 93c, respectively. The two largest fossil energy inputs in the wood cellulose production system were steam and electricity. Note that 25% lignin was credited in this system, as it was in the switchgrass calculations.

Biodiesel Production

The monoesters commonly known as biodiesel are usually produced through the transesterification of vegetable oils (triglycerides or fatty acids) with the fossil methane-derived methanol. Both liquid oils and solid fats are triglycerides or fatty esters of glycerin. Raw triglyceride vegetable oils have properties similar to those of petroleum-based diesel fuels and can be used as a direct replacement without engine modification for a limited operating duration. Long-term diesel operation on raw vegetable oils causes numerous problems including injector coking, contamination of lubrication oil, engine deposits, and increased emissions. These problems are primarily the result of the high viscosity of the triglyceride oils compared to petroleum-based diesel fuels.

The biodiesel recipe is 100 kg of soybean oil, 15 kg of methanol, 1 kg of catalyst (sodium hydroxide or lye). The products are 100 kg biodiesel, 10 kg glycerol, 5 kg methanol (reusable after separation and purification) and 1 kg of soaps. The lye catalyst is neutralized with hydrochloric or sulfuric acid and becomes a waste product.

Soybean Conversion into Biodiesel

Various vegetable oils have been converted into biodiesel and they work well in diesel engines. An assessment of producing sunflower oil proved to be energy negative and costly in terms of dollars (Pimentel, 2001). Although soybeans contain less oil than sunflower, about 18% soy oil compared with 26% oil for sunflower, soybeans can be produced without or nearly zero nitrogen. This makes soybeans advantageous for the production of biodiesel. Nitrogen fertilizer is one of the most energy costly inputs in crop production (Pimentel and Patzek, 2005). The yield of sunflower is also lower than soybeans, about 1,500 kg/ha for sunflower compared with 2,890 kg/ha for soybeans (USDA, 2004). The production of 2,890 kg/ha of soy requires an input of about 3.0 million kcal per hectare and costs about $473/ha.

With a yield of oil of 18% then 5,556 kg of soybeans are required to produce 1,000 kg of oil. The production of the soy feedstock requires an input of 5.7 million kcal. The second largest input is steam that requires an input of 1.4 million kcal. The total input for the 1,000 kg of soy oil is 13.8 million kcal. In addition, 125 kg of methanol must be added to produce biodiesel fuel. The methanol has an energy value of 587,500 kcal. With soy oil having an energy value of 9 million kcal, then there is a net loss of 53% in energy.

A credit should be taken for the soy meal that is produced; this has an energy value of 7.4 million kcal, but it must be emphasized that this soy meal is not liquid fuel but livestock feed. The price per kilogram of soy biodiesel is about $1.12. Note, soy oil has a specific gravity of about 0.92; thus soy biodiesel value per liter is 97c per liter. This makes soy oil about 1.8 times more expensive than diesel fuel; diesel costs about 53c per liter to produce (USCB, 2004–2005). Soybeans are a valuable crop in the United States.

Canola Conversion into Biodiesel

Another crop that can be converted into biodiesel is canola that produces the most valuable cooking oil. Although soybeans contain less oil than canola, about 18% soy oil compared with 30% oil for canola, soybeans can be produced without or nearly zero nitrogen. This makes soybeans advantageous for the production of biodiesel. Nitrogen fertilizer is one of the most energy costly inputs in crop production (Pimentel and Patzek, 2005).

The yield of canola is also lower than soybeans, about 1,600 kg/ha for canola compared with 2,890 kg/ha for soybeans (USDA, 2004). The production of 1,568 kg/ha canola requires an input of about 4.4 million kcal per hectare and costs about $573/ha.

About 3,333 kg of canola oil is required to produce 1,000 kg of biodiesel. The total energy input to produce the 1,000 of canola oil is 14 million kcal. This suggests a net loss of 58%. The cost per kg of biodiesel is also high at $1.63.

CONCLUSION

Several physical and chemical factors limit the production of biofuels such as ethanol and biodiesel from plant biomass. Fossil energy inputs needed in the production of ethanol from corn or cellulosic wood material are several times more than the ethanol energy output. For biodiesel produced from soybeans, fossil energy inputs are 40% greater than contained in the biodiesel fuel produced. Giving credit for the byproducts produced can reduce the fossil ene gy inputs only from 10% to 20%. An extremely low fraction of the sunlight reaching a hectare of cropland is captured by green plant biomass. On average only 0.1% of the sunlight is captured by plants. This value is in sharp contrast to photovoltaics that capture more than 10% of the sunlight, or approximately 100–fold more sunlight than the green plant biomass. The environmental impacts of producing either ethanol or biodiesel from biomass are enormous. These include: severe soil erosion; heavy use of nitrogen fertilizer; and use of large quantities of pesticides (insecticides and herbicides). In addition to a significant contribution to global warming, there is the use of 1,000–2,000 liters of water required for the production of each liter of either ethanol or biodiesel. Furthermore, for every liter of ethanol produced there are 6–12 liters of sewage effluent produced.

Burning food crops, such as corn and soybeans, to produce biofuels, creates major ethical concerns. More than 3.7 billion humans are now malnourished in the world and the need for food is critical. Energy conservation strategies combined with active development of renewable energy sources, such as solar cells and solar-based methanol synthesis systems, should be given priority.

13

Developing Energy Crops for Thermal Applications

The main historic constraint in the development of herbaceous biomass for thermal applications has been clinker formation and corrosion in the boiler during combustion. This problem is being overcome through plant selection and cultural techniques in grass cultivation, combined with advances in combustion technology. In the coming years, growing warm-season grasses for pellet production will emerge as a major new renewable energy technology, largely because it represents the most resource-efficient strategy to use farmland in temperate regions to create energy security and mitigate greenhouse gases.

INTRODUCTION

In most industrialized countries, thermal energy represents the largest energy need in the economy. Thermal energy is used for space and water heating in the residential, commercial and industrial sectors, low and high temperature process heat for industry, and power applications. Thermal energy can also be used for cooling applications. Rather than supporting biomass for simple thermal applications such as direct heating applications industrialized countries have currently placed emphasis on researching and providing subsidies for more technologically complex innovations such as large industrial bio-refineries. However, governments in industrialized nations who have identified the need to develop biofuels for energy security and greenhouse gas mitigation should look more closely at thermal applications for biomass to fulfill these needs. This review therefore examines energy security in section one, identifying opportunities to grow energy crops on farmland in eastern Canada as a means to collect solar energy and convert it into useful energy products for consumption. The greenhouse gas (GHG) mitigation potential of switching from fossil fuels to various biofuels produced from energy crops is also examined.

ENERGY CROP PRODUCTION FOR ENERGY SECURITY AND GHG MITIGATION

Since the Arab oil embargo in the 1970s there has been considerable interest in North America in growing both conventional field crops and dedicated energy crops for bioenergy as a means to enhance energy security. The long-term decline in farm commodity prices has also created significant interest in using the surplus production capacity of the farm sector as a means to produce energy while creating demand enhancement for the farm sector. This decline in farm commodity prices, due to innovation in plant breeding and production technology, is accelerating the likelihood that large quantities of biomass energy from farms could penetrate energy markets currently dominated by fossil fuels.

One of the strongest drivers for biofuel development is the GHG mitigation potential of energy crops to produce solid, liquid and gaseous biofuels to replace fossil fuels in our economy. With the increased use of grain crops for liquid biofuels, the past two years have seen a rise in both the demand and price for farm commodities. Also increasing however, are concerns over other important social issues such as the potential for bioenergy to compete with food security, and problems with soil erosion and long-term soil fertility. The production and utilization of crops residues as a global biofuel sources has recently been reviewed (Lal, 2005). The main conclusions were that the most appropriate use of crop residues is to enhance, maintain and sustain soil quality by increasing soil organic matter, enhancing activity and species of soil fauna, minimizing soil erosion and non-source pollution, mitigating climate change by sequestering carbon in the pedosphere, and advancing global food security through enhancement of soil quality. It was recommended that efforts be undertaken to grow biomass on specifically dedicated land with species of high yield potential, suggesting that 250 million hectares (ha) globally could be put into production of perennial energy crops.

The increasing biodiversity loss from agricultural landscapes through crop intensification is also a major environmental concern. The rapid development of liquid biofuels in the tropics in the past decade has also caused significant harm to biodiversity through the conversion of forests into agricultural production. Resource efficient, rather than resource exhausting, bioenergy crop production strategies need to evolve with a priority placed on de-intensification of farm production through the use of perennials and utilization of existing marginal farmlands. This approach would to a much greater extent avoid the biofuel conflicts with food crop production and biodiversity that are now occurring with using annual food crops as biofuels.

To achieve the objective of resource efficient biomass production we must examine some of the basic factors influencing biomass accumulation:

- There are two main photosynthetic pathways for converting solar energy into plant material: the C3 and C4 pathways. The C4 pathway is approximately 40% more efficient than the C3 pathway in accumulating carbon (Beadle and Long, 1985).
- C4 species use approximately half the water of most C3 species (Black, 1971).
- In temperate climates, sunlight interception is often more efficient with perennial plants because annual plants spend much of the spring establishing a canopy and also exhibit poor growth on marginal soils.
- Some species of warm season grasses are climax community species and have excellent stand longevity (which also results in decreased economic costs for establishing perennial crops through decreased expenditures for seeding, tillage etc.).
- C4 species of grasses contain less N than C3 species and can be more N-use efficient in temperate zones because

the N is cycled internally to the root system in the fall for use in the following growing season (Clark, 1977).

- It is apparent that the optimal plants for resource-efficient biomass production should be both perennial and C4 in nature.

PERENNIAL AND ANNUAL ENERGY CROPS

In North America, the warm continental climate has produced a diversity of native warm season (C4) perennial grasses that have a relatively high energy production potential on marginal farmlands. In the more humid zones, these species include switchgrass (*panicum virgatum*), prairie cordgrass (*spartina pectinata*), eastern gamagrass (*tripsacum dactyloides*), big bluestem (*andropogon gerardii vitman*) and coastal panic grass (*panicum amarum A.S. hitchc.*). In semi-arid zones and dry-land farming areas, prairie sandreed (*calamovilfa longifolia*) and sand bluestem (*andropogon hallii*) are amongst the most productive species. All of these species are relatively thin stemmed, winter hardy, highly productive and are established through seed.

Switchgrass was chosen as the model herbaceous energy crop species to concentrate development efforts on in the early 1990s by the U.S. Department of Energy. It had a number of promising features including its moderate to high productivity, adaptation to marginal farmlands, drought resistance, stand longevity, low nitrogen requirements and resistance to pests and diseases (Samson and Omielan, 1994; Parrish and Fike, 2005).

The perennial crops were also identified to have the lowest fossil energy input requirements. Overall, prior to any conversion process, switchgrass produces 40% more net-energy gain per hectare than grain corn and five times more net-energy gain per hectare than canola. It also should be noted that corn yields are based on modern hybrid yields in Ontario while switchgrass yields are based on commercial production of the cultivar cave in-rock, an unimproved cultivar that was collected from an Illinois prairie in 1958. Warm season grasses (WSG's) function well as perennial energy crops because they mimic the biological efficiency of the tall-grass prairie ecosystem native to North America. They produce significantly more energy than grain corn while at the same time requiring minimal fossil energy inputs for field operations and less fertilizers and herbicides.

In industrialized countries, the seed portion of annual grain and oilseed crops became the first feedstock for energy applications. However, whole plant annual crops capture much larger quantities of energy per hectare. InWestern Europe, whole plant crops such as maize and rye are now commonly harvested for biogas applications. High yielding hybrid forage sorghum, sorghum-sudangrass and millet, also hold promise as new candidates for biogas digestion (Von Felde, 2007; Venuto, 2007). The major advantage of ensiling is that even in relatively unfavourable weather for crop drying, energy crops can be stored and delivered to the digester year round. This is particularly advantageous for thick stemmed species like maize and sorghum which are commonly difficult to dry in areas receiving more than 700mm of rainfall annually or have harvests late in the year when solar radiation is declining. In combustion applications, thick stemmed herbaceous species have biomass quality constraints which make them difficult to burn. In warm, humid southern production zones in temperate regions, it may also be difficult to dry the feedstock for combustion applications as the material would be more vulnerable to decomposition. In these situations, crop conversion to usable energy would be facilitated by using a biogas conversion system and storing the crop as silage. Overall, both thick and thin stemmed whole-plant biomass crops can be successfully grown for biogas applications. Highest biogas yields are achieved when a fine chop and highly digestible silage are used. Conversely, thin stemmed, perennial WSG's have been identified as the most viable means to store dry crops for combustion applications and offer the best potential for improved biomass quality for combustion. For liquid fuel production such as cellulosic ethanol, the process is more flexible in terms of the moisture content and chemical composition of the feedstock in the production of energy.

Options for Growing and Using Energy Crops for Energy Security in Industrialized Countries

As greater scarcity of fossil fuels occurs in the next 25–50 years, industrialized countries will undoubtedly seek greater energy security from renewable energy technologies (RET's).

Countries will increasingly aim to develop bioenergy production and conversion technologies which are efficient at using energy crops grown on both productive and marginal farmland to displace the use of imported fossil fuels. North America, Europe, and China in particular, urgently need to develop effective bioenergy production systems as these areas will become increasingly dependent on importing fossil fuels due to their large economies and declining fossil energy production. While many industrialized countries have imported petroleum fuels from distant producers for many years, the international trade in natural gas use will expand substantially. For example in North America, domestic natural gas production peaked in the United States in 2001 and has declined by 1.7% per year since that time, while in Canada production has been in decline or reached a plateau since 2001. To compensate for declining North American gas production and rising prices, energy intensive natural gas industries have moved offshore and liquid natural gas (LNG) imports have started to come into the United States (Hughes, 2006). LNG imports currently supply approximately 3% of the United States supply and are expected to increase to 15–20% by 2025. Much of this natural gas demand is presently used in thermal applications. For example, the United States relies on natural gas for 20% of its power requirements and for 60% of its home heating requirements (Darley, 2004).

Identifying sustainable bioenergy technologies with a high net energy gain per hectare is essential to reduce imports of natural gas and other fossil fuels into industrialized countries. In particular, there may be opportunities to cost-effectively produce solid and gaseous biofuels in temperate regions to replace high quality fossil fuels in thermal applications. In the past 5 years, petroleum and natural gas prices have increased substantially while thermal coal prices in the world have remained relatively stable. This likely is a function of the changing awareness around supply and demand of fossil fuels. On a global basis, the lifespan of natural gas and oil reserves are less than half that of coal, however many energy analysts foresee a transition from the current global energy economy dominated by petroleum to one where natural gas plays an equally important role. This widening gap between the prices of high-quality fossil fuels like natural gas and petroleum versus coal will make fuels of higher quality ideal candidates for displacement by renewables. Solid and gaseous biofuels could substitute in thermal applications through both heat generation and combined heat and power operations. This is a fitting association as both biomass production and heat demand are relatively disperse, thus biomass could be produced locally to meet local thermal energy needs sustainably. A key tenet of the concept of the *soft energy path* introduced by Lovins (1977) is that both the scale and quality of energy should be matched appropriately with its end use to create a more sustainable energy supply system.

The growing price difference between coal, natural gas and heating oil suggests that high-quality fossil fuels will be increasingly utilized for high-quality end uses such as transportation fuels and industrial products while lower-quality fuels like coal will be increasingly used for low-end thermal applications. Due to the polluting nature of coal and the increasing emphasis on reducing carbon emissions through taxes and cap and trade systems, there also will be substantial opportunities for biomass to substitute for coal in thermal applications. The following section explores the thermodynamics around converting biomass into solid and gaseous products versus their present utilization opportunities as liquid fuels in temperate regions of the world.

Opportunities to use Ontario Farmland for Improving Energy Security

This analysis examines present or currently proposed strategies to use biomass derived from farmland in the province of Ontario for generating solid, gaseous and liquid biofuel products. Ontario has a continental climate and cropping patterns that are somewhat similar to other regions in the temperate world including the Great Lake states of Michigan and Wisconsin in the United States, countries in central Europe such as Hungary, and the Northeastern provinces of China. As such, it represents a useful case study for the bioenergy opportunities for continental climates in the temperate world. Ontario produces very limited quantities of fossil fuels. Coal and coal products in Ontario are primarily used for power generation and for large industrial applications, such as the steel and cement industry. Petroleum products are mainly used in the transport sector in Ontario, with some additional use as heating oil. Ontario imports natural gas from western Canada, petroleum from the world market, and coal mainly from the Northeastern United States. Within the next 2–5 years, two LNG terminals on Canada's east coast will begin supplying eastern Canadian energy user's imported liquefied natural gas from either Russia or producers in the Middle East. Declining western Canadian supplies will likely not be sufficient to enable export production to reach Ontario in the coming years. Thus the Ontario economy, which is heavily dependent on natural gas for residential and commercial heating applications and process heat for industry and power generation, will begin to rely on distant foreign natural gas resources.

Harvesting Energy from Ontario Farmland for Biofuel Applications: A Case Study

To optimize energy security and GHG mitigation potential from bioenergy, a case study has been developed to compare alternative bioenergy crops and conversion technologies in Ontario. The comparison crops include soybean, corn, corn silage and switchgrass, which are well adapted to Ontario's warm continental climate summer. The main agricultural zones in the province experience a frost free period typically from mid May to mid to late Sept and about 900mm of annual precipitation. Soybean, corn and corn silage are commonly grown in Ontario while switchgrass and other native warm season grasses such as big bluestem and coastal panic grass are emerging crops that are native to the region. Switchgrass has been selected to represent the WSG's in the analysis as it has undergone the furthest development of all native grasses for energy use in North America. In Ontario, approximately 500 ha of native grasses are presently under bioenergy production in 2007. It is anticipated that a portfolio of warm season species will be developed as future energy crops, with mixed seedings encouraged to reduce production risks and enhance biodiversity.

The dry matter production potential prior to processing is highest with the whole corn plant harvested as silage. Switchgrass also produces significant quantities of dry matter and has the added advantage of being able to be grown on marginal farmlands.

The yields for switchgrass are estimated to be slightly lower for combustion applications as a delayed harvest technique is used. The net energy gain/ha that results from each energy crop and conversion process is generally highest where whole-plant biomass is used for biogas or bioheat, and lowest where the seed portion of annual crops is used for liquid fuels.

From a net energy gain perspective, the two most promising systems for Ontario are corn silage biogas and switchgrass pellets. These technologies have the potential to produce 770–890% more net energy gain/ha than growing grain corn for ethanol. Cellulosic ethanol from grasses is much more efficient than other annual grain or oilseed liquid fuel options for producing net energy gain/ha. However it remains substantially less efficient than direct combustion of energy grasses or corn silage biogas as a means to produce energy from farmland.

The main problem of cellulosic ethanol is that, even with current technology, less than half of the energy in the original feedstock is recovered in the ethanol. This analysis illustrates that upgrading the energy quality of biomass from a solid form to a liquid form appears to be quite expensive thermodynamically. While advances in cellulosic ethanol technology can be expected in the coming years, the prediction of a technology that would be cost-competitive at $1.00/gallon with gasoline by the year 2000 (Lynd et al., 1991), was and remains far from reality. There are currently no commercial cellulosic ethanol plants using agricultural feedstocks in existence despite the generous subsidies for ethanol production available in North America.

Biogas production from energy crops represents a more thermodynamically efficient option than converting plant matter into liquid fuels. Considering the case of corn silage, 500m3 of biogas can be produced from one tonne of feedstock which is equivalent to 11.6 GJ/ODT or 61.7% conversion efficiency. In contrast with current projected cellulosic ethanol product yields of 340 l of ethanol (Iogen Corporation, 2008), 7.1GJ/tonne of energy is recovered, a 38% conversion efficiency. In Germany, there has been significant scale-up of energy crops grown for biogas applications. In 2006, there were an estimated 3500 biogas digestors in the country that were mainly operating on energy crops such as corn silage, rye silage, and perennial grasses as well as manure and food processing wastes (House et al., 2007). Some of the main problems facing the cellulosic ethanol industry are: (1) a chronic underestimation of feedstock procurement costs required by farmers in industrialized countries to make the technology viable on a large-scale; and (2) projected commercial plant construction costs have risen dramatically, especially those for stainless steel and skilled labour costs.

The economics now favour larger plants, with most plants foreseen to have a feedstock requirement of one million tonnes per year or more. Considering the increasing depletion of fossil energy resources in industrialized countries projected for the future, it may be difficult for such large amounts of affordable biomass to be procured and transported to bioethanol plants, especially when they are competing with local biogas and bioheat plants that are more thermodynamically efficient and have significantly lower processing and transport costs.

A centralized biogas digester producing 3MW of thermal energy or a 50,000 tonne per year bioheat pellet plant have much smaller land area footprints than a 700,000 tonne per year cellulosic ethanol plant. The land area to be planted, based on the switchgrass yields, would be 6000ha and 75,000ha for a switchgrass pellet and cellulosic ethanol plant respectively.

If 1 in 4 ha surrounding the cellulosic ethanol plant was planted to switchgrass, the plants feedstock supply would be drawn from a land area covering 300,000 ha and stretch out a radius of 310km from the plant. The economic premium offered to produce liquid biofuels as a substitute for gasoline may not be sufficient to recover the large thermodynamic loss required for production and conversion of solid plant matter into liquid fuel in these large biorefineries.

It is, by comparison, more efficient to use whole-plant biomass in pellet or biogas form to substitute for natural gas. As such, in temperate regions of industrialized countries, which are densely inhabited and have high local demands for heat and power, bioheat and biogas are the technologies likely to succeed if there is any level of parity in the government incentives applied to bioheat, biogas and liquid biofuels.

Greenhouse Gas Mitigation from Bioheat and Other Biofuels Options

With increasing concern about global climate change, it is of paramount importance that cost-effective emission reduction strategies evolve from producing bioenergy from farmland in industrialized countries. Efforts to import biofuels from tropical countries to date have resulted in rapid deforestation of native forests for palm oil production, particularly in Malaysia and Indonesia. Sugar cane cultivation for ethanol production is now expanding into traditional grazing lands in countries like Brazil, causing the cattle industry to expand into tropical forests. While certification systems may evolve for sustainable importation of tropical biofuels into industrialized countries, it is essential that effective domestic strategies are developed in industrialized countries to reduce the need for these imports. Developing nations in the tropics will themselves require large volumes of biofuels for their internal needs, further increasing the pressure on industrialized nations to become energy self-sufficient.

An important driver for the development of bioenergy will be the economic competitiveness of various technologies as greenhouse gas mitigation strategies. Thus, it is important that the economics of various solid and liquid biofuels options be compared. There are several factors which are fundamental to the economic competitiveness of various agricultural biomass production and conversion chains to reduce greenhouse gases effectively including:

1. optimizing the amount of energy produced from each hectare of marginal and arable farmland;
2. the net GHG offset provided by displacing a GJ of fossil fuels used for a particular application with a renewable energy for the same application (fuel switching); and
3. the cost of production of the processed bioenergy product relative to the fossil fuel it is displacing.

The net offsets from various bioenergy technologies through fuel switching applications for fossil fuels in Ontario, Canada. The net GHG offsets are highest with switchgrass pellets (86–91%), moderate with soybean biodiesel (50% offset) and low with corn ethanol (21%). The low GHG offsets from corn ethanol is confirmed by two recent analyses in the United States which determined the GHG offset potential of corn ethanol to be 15% (Farrell et al., 2006) and 19% (Wang et al., 2007) respectively in the current state of the industry.

The reason for the high offset potential of switchgrass pellets is that they require modest amounts of energy for switchgrass feedstock production and pellet processing. As well there is no change of physical state that occurs, so nearly all of the energy content of the grass is available in a pellet form.

Switchgrass production also has no significant landscape emissions as N2O emissions are low for perennial grasses and the soil carbon sequestered is expected to offset the low amounts of N2O emissions that occur (Adler et al., 2006).

Soybean biodiesel (or canola biodiesel) represents a moderately efficient offset potential because the liquid fuel production process is not energy intensive and the crop has moderate energy inputs and N2O emissions in North America (Samson et al., 2008a).

Each GJ of soybean biodiesel produced displaces approximately half the GHG emissions of diesel fuel. However it still represents a largely ineffective approach to mitigate greenhouse gasses from farmland in temperate regions, as the soybean yield is low and the oil content in the soybean seed is low. With soybean oil being a high quality vegetable oil selling for premium prices in 2007–2008 around $1000/tonne, biodiesel is far from being an economically viable biofuel unless heavily subsidized.

The reasons for the low offset potential of corn are: (1) the technology relies heavily on carbon intensive fuels such as coal and natural gas for processing; (2) corn is an energy intensive annual crop to produce; (3) there are relatively high N2O losses from each hectare of corn production, which has a strong impact on overall emissions; and (4) comparatively low amounts of energy are captured in the field and converted into a final energy product. In the province of Ontario, Canada, the combined federal and provincial incentives in 2007 amounted to 16.8 cents per litre of ethanol produced (or $8.00 CAN/GJ assuming an energy value of 0.021GJ/litre). With only 21.13 kg CO2e offsets per GJ of fuel, it takes 47.3 GJ of ethanol to offset one tonne of carbon dioxide. This is equivalent to a subsidy of $379 (CAN) (Samson et al., 2008a) (1 $CAN = $1 USD in October 2007). Even larger federal and state subsidies are available for promoting corn ethanol production in certain states in the United States.

The main advantage of cellulosic ethanol from switchgrass over corn ethanol is that the heat and power for plant processes are provided by lignin, a by-product in cellulosic ethanol processing.

Cellulosic ethanol results in a moderately high offset potential of 76.5% compared to the use of gasoline. Nevertheless this use of lignin causes a parasitic impact on the net GHG mitigation per ha that can be provided in comparison to using the grass for other bioenergy applications.With relatively modest volumes of energy recovered from each tonne of biomass of 340 l ethanol/tonne, the technology on a per hectare basis represents only a moderately efficient approach at using farmland to mitigate greenhouse gases.

The technology can be best categorized as having a medium energy output per hectare and a moderate to high GHG offset when displacing fossil fuels. It is significantly superior to corn ethanol and soybean biodiesel if current commercialization problems can be overcome.

It can be observed that corn ethanol and soybean biodiesel cannot be considered effective greenhouse gas mitigation policies with less than 1,500 kg CO2e/ha offsets.

Per hectare, corn ethanol has modest energy production and poor net GHG offsets, while per hectare soybean biodiesel has a poor liquid fuel output and only a moderate GHG offset. This analysis demonstrates that solid biofuels represent a highly promising means for Ontario to mitigate greenhouse gases, particularly compared with liquid fuel options.

The advanced boiler technology currently available to burn pellets offers the same combustion efficiency as natural gas combustion appliances (Fiedler, 2004). When switchgrass pellets are used to displace coal, the highest overall GHG displacement potential can be achieved at 13,098 kg CO2e/ha. The lowest GHG emission potential of switchgrass pellets is at 7,617 kg CO2e/ha when they are used to displace natural gas.

When switchgrass pellets replace imported LNG from Russia, approximately 10 tonnes of CO2e/ha is abated. From Ontario's perspective, an effective policy strategy for GHG mitigation would clearly be to replace foreign imports of LNG and coal with domestically produced pellets within the province.

A $2(CAN)/GJ incentive for switchgrass pellet producers would cost an average of $24, $31 and $40/tonne of CO_2 offset to displace the use of coal, liquefied natural gas and conventional gas, respectively. In contrast, Ontario has combined federal and provincial wi..d energy incentives of $15.28/GJ (6.5 cents/kWh), soybean biodiesel incentives of $5.68/GJ (20 cents/L) and corn ethanol incentives of $8.00/GJ (16.8 cents/litre).

The corresponding costs of these offsets are $50, $98 and $379/tonne CO_2e for wind, biodiesel and corn ethanol, respectively (Samson et al., 2008a). Two other recent studies also found that with carbon taxes under $100/tonne, bioheat is considerably less expensive GHG offset strategy than producing liquid fuels in temperate regions (Grahn et al., 2007). To create more effective use of taxpayers' money in reducing GHG emissions, policy makers need to understand the offset potential of the various technologies and create mechanisms to allow GHG reduction to happen competitively within the marketplace.

Another problematic example exists with energy crop use for biogas systems, which is currently strongly supported as a RET in Germany. Energy crop biogas systems appear to be facing several challenges in being an efficient GHG mitigation technology. Few detailed studies have been completed but there appears to be some identified limitations.

When examining only energy related GHG emissions, power generation from energy crop biogas is a highly effective GHG mitigation technology compared to using fossil fuels for power production (Gerin et al., 2008). However, two GHG emission problems have been identified with energy crop biogas for power generation which are methane leakage from digesters (estimated at 1%) and the high N2O emissions associated with maize cultivation (Crutzen et al., 2007). In one preliminary study from Western Europe there was no net GHG benefit from maize silage biogas because of these two aforementioned problems (Zwart et al., 2007).

It is likely the use of deeper rooted and more nitrogen efficient annual crops such as sorghum or perennial species such as highly digestible warm season grasses may help reduce GHG emissions from feedstock production.

As well, energy crop cultivation in less humid regions would reduce the N_2O loss problem. New design features of digesters and larger centralized biogas digesters may help reduce methane losses that are currently occurring. Energy crops used in biogas digesters in the future will likely play an important role in providing GHG friendly thermal energy for combined heat and power applications.

However, presently only manure biogas digesters have been found to have positive impacts on GHG mitigation (Zwart et al., 2007). If governments created incentives for RET's based on their actual GHG mitigation efficient approaches to reduce emissions would be likely be stimulated and more efficient progress in mitigating GHG's would be realized through bioenergy technologies.

OPTIMIZATION OF ENERGY GRASSES FOR COMBUSTION APPLICATIONS

From the previous analysis it is evident growing energy grasses for bioheat represents the most outstanding option for using one hectare of farmland to produce renewable energy and mitigate GHG's from an agricultural production system. If energy crop grasses are to evolve as a major new RET for energy security and GHG abatement for the industrialized world, it is imperative that considerable research and development efforts to expand this opportunity be undertaken.

Historically, the major limitation to the development of grasses for bioheat applications has been the difficultly associated with burning energy grasses efficiently in conventional biomass boilers.

In particular, the relatively high alkali and chlorine contents of herbaceous plants are widely known to lead to clinker formation and corrosion of boilers. These biomass quality problems have resulted in slow commercialization of grass feedstocks as agro-pellets for use in small scale boilers (Elbersen et al., 2002; Obernberger and Thek, 2004).

Despite this, the problems with burning grasses have now become reasonably well understood and constraints are being resolved through several strategies. Plant selection and breeding together with delayed harvest management can be used to reduce the chlorine, alkali and silica content in native grasses, reducing clinker formation and corrosion in boilers.

Utilizing advanced combustion systems which are specifically designed to burn high-ash; herbaceous fuels can also reduce problems with ash accumulation in burners (Obernberger and Thek, 2004). However, high-ash fuels can still pose major convenience issues, particularly when used in pellet stoves and small scale boilers. Strategies to lower the ash content and the undesirable chemical elements in grasses are essential if commercial markets are to be fully developed.

Improving Biomass Quality for Combustion

The most serious biomass quality problem with herbaceous feedstocks is the alkali and chlorine content in the feedstock material, which has potential for fouling and corroding boilers during combustion (Passalacqua et al., 2004).

Particulate emissions are strongly related to fuel type, and specifically, to the content of aerosolforming compounds such as potassium (K), chlorine (Cl), sodium (Na), sulphur (S) and even lead and zinc in the fuel (Hartmann et al., 2007).

Using fuels that are low in the "dust critical" elements K, Cl, Na and S is of particular importance for achieving high-quality biomass fuels and lowering particulate emissions during biomass combustion.

The major factors affecting the level of aerosol-forming compounds are fertilization practices, choice of species, stem thickness, time of crop harvest, relative maturity of the cultivar, and the level of precipitation in a region (Samson et al., 2005; Samson, 2007).

Chlorine is particularly problematic as it increases the ash-sintering effect of fuels containing potassium and makes these elements migrate from the fuel bed to the boiler walls, forming clinkers (Godoy and Chen, 2004).

The nitrogen content of feedstocks has little impact on the efficiency of the combustion process but burning high-N fuels is undesirable from an environmental standpoint as this contributes to NOx pollution. However, delayed harvest switchgrass has relatively low N contents that are comparable to wood (Samson et al., 2005; Adler et al., 2006). Reducing the moisture content of feedstocks to below 15% is also important as this eases storage problems from decomposition and can reduce or even eliminate the need to dry materials before pelletizing them.

NUTRIENT MANAGEMENT

Both potassium and chlorine are known to be effectively leached out of thinstemmed grasses in humid climates. As potassium is water soluble, the potassium content in plants can decrease appreciably following senescence of materials during the end of growing season, particularly if significant rainfall occurs during this period.

Prairie ecology studies have also demonstrated that potassium in unharvested material is efficiently recycled into the soil over the late fall and winter (Koelling and Kucera, 1965; White, 1973). Kucera and Ehrenreich, (1962) in Missouri found potassium content of native prairie plants to decline from 1.34% K2O in mid-June, to 0.63% by mid-September, and to 0.05% by the end of November. Koelling and Kucera (1965) found the average potassium content of big bluestem in the Missouri prairies to decrease from 1.28% K2O in July, to 0.33% in September, and to 0.13% in November. Over-wintering further reduced levels to 0.07% by May the following year.

It is also of interest to note that native prairie materials likely have significantly earlier maturity dates (and hence time for fall leaching) than purpose grown energy grasses. In Quebec, Cave-in-Rock switchgrass harvested in early October was found to contain 0.95% potassium, while over-wintered switchgrass harvested in mid-May was found to contain just 0.06% potassium (Goel et al., 2000).

In the case of potassium, it appears that harvesting in the fall at least several weeks after materials senesce, or alternately harvesting over-wintered material, provides significant reductions in the potassium content of feedstocks. Chlorine is also highly water soluble in herbaceous biomass feedstocks (Sander, 1997). Like potassium, the chlorine content of perennial grass feedstocks is reduced if a late-season or overwintering harvest management regime is practiced. Burvall (1997) found an 86% reduction in chlorine content of reed canarygrass when it was over-wintered in Sweden.

HARVEST MANAGEMENT AND CULTIVAR SELECTION

Despite the benefits that overwintering can provide, letting grasses remain unharvested through the winter can also reduce the eventual biomass yield obtained in the spring. In Southwestern Quebec, spring-harvested switchgrass yields were found to be approximately 24% lower than that of fall-harvested switchgrass (Goel et al., 2000).

This loss was due likely to both the late season translocation of materials to the root system in winter (Parrish et al., 2003), and the physical loss of material, mainly from leaves and seed heads during the winter season (Goel et al., 2000). Compared to fall harvested material, spring-harvested switchgrass lost 4% of dry matter from the stem component, 11% from leaf sheaths, 30% from leaves and 80% from seed heads (Goel et al., 2000). Field observations have indicated that when the material is completely dry in late winter and early spring, the majority of breakage losses occur during storm events. As well, some decomposition occurs in the field when material lodges in late summer and early fall and plants come into contact with the soil.

A new delayed harvest technique was assessed in the spring of 2007 in Ontario by REAP-Canada (Samson et al., 2008b) to minimize winter breakage and spring harvest losses from feedstocks, while maintaining the benefits of nutrient leaching that are associated with overwintering. Under this system, the material is mowed into windrows in mid-November and directly baled off the windrow in the spring. Results to date are promising as yields were 21% higher than spring mowed and harvested material. The fall mowing technique also caused faster spring drying of windrowed material, but recovery of material below 10% moisture was achieved in early May in both systems. Finally, the fall mowing technique encouraged earlier soil warming and than spring mowed areas, promoting earlier regrowth of the switchgrass.

Selecting for increased stem and leaf sheath content and developing warm season grass varieties that more efficiently retain their leaves through the winter could help reduce overwintering losses. Another strategy that has proven effective to reduce potassium and chlorine content in feedstocks is to utilize earlier-maturing warm season grass varieties that senesce earlier in the fall (Bakker and Elbersen, 2005).

Early maturity enables a more extended period between senescence and late fall harvest for nutrients to be leached from the stem material. Thin stemmed grasses have also been identified to have higher nutrient leaching potential compared than thicker stemmed grasses.

Lowland switchgrass cultivars with tall, moderately coarse stems, such as Alamo and Kanlow have been found to be moderately higher in K and Cl than upland switchgrass with short, fine stems at the end of the season (Cassida et al., 2005). The average outer diameter of lowland ecotypes of switchgrass has been found to range from 3.5mm (Igathinathane et al., 2007) to 5mm (Das et al., 2004) and to have a stem wall thickness of approximately 0.7mm (Igathinathane et al., 2007).

The problem of biomass quality appears to be even more serious in miscanthus than switchgrass. Thick stemmed miscanthus ecotypes are known to have high potassium and chlorine contents, especially when combined with late maturity (Jørgensen, 1997). Comparatively, the average stem diameter of miscanthus is 8.8–9.2 mm, with a stem wall thickness of 1.3–1.5mm (Kaack and

Schwarz, 2001). No biomass quality reports from Europe could be identified which indicated miscanthus sinensis giganteus was able to reach the minimal biomass quality targets of 0.2% K and 0.1% chlorine for power generation in Denmark outlined by Sander (1997). Thick stems also make it more difficult to dry material.

REAP-Canada identified that even in fall harvested upland switchgrass, while most plant components had moisture contents below 15%, the stems still tended to retain significant moisture (Samson et al., 2008b). Spring harvesting of material can enable bales to be collected below 12% moisture.

The low moisture content of grasses at spring harvest is a significant advantage that grass energy crops hold over woody energy crops. The moisture content of willows at harvest for willows can be 50%. High moisture woody materials can use 21% of the raw material to provide energy for the drying process if made into pellets (Bradley, 2006). Spring harvested grasses thus have a major biomass quality advantage for pellet processing because of the dryness of the material.

Overall, grass pellets appear to represent the most promising solution to the strong international growth in demand for fuel pellets, a growth that cannot be met with supplies of wood residues forecast for the future.

Many combustion issues have now been resolved in replacing wood pellets with grass pellets. Research indicates native warm-season grass pellets grown in North-eastern North America can approach a comparable content of aerosol forming compounds as that found wood residue pellets. However, the overall ash content of grass pellets typically remains considerably higher than wood. Wood residue pellets of highest quality are sold as premium grade when they achieve less than 1% ash. Typically, the European market trades wood pellets with 0.6% ash in this category (Obernberger and Thek, 2004). However, grasses harvested in North-eastern North America are generally in the 3- 5% ash range (Samson et al., 2005). Even higher contents of ash are experienced in switchgrass growing regions with less favourable rainfall to evaporation ratios such as western Canada (Jefferson et al., 2004) and the Western United States (Cassida et al., 2005).

Impacts and Management of Silica

Silica levels in grasses must also be reduced if grass pellets are to enter into the high-end residential wood pellet market that currently has products trading in Europe at approximately $250/ tonne.

Producing fuels with lower silica levels has many benefits. Low silica containing fuels have higher energy contents, reduce abrasion on metal parts such as pellet dies during the densification processes, and improve convenience in reducing ash removal requirements.

When burned in pellet appliances, high-ash grass pellets with high silica contents can also produce a low-density ash that retains the shape of the former pellet. As an example, consider that the bulk density of reed canary grass ash has been assessed to be half that of wood ash (Paulrud, 2004).

Thus the residual ash leftover after burning grass pellets in the 3–5% ash range can take up to 10–20 times the volume of the ash from burning 0.6% ash wood pellets. To burn 3–5% ash grass pellets, ash pans will need to be modified in smaller appliances to create larger ash collecting areas.

Combustion units burning high-ash grass pellets will require more frequent cleaning and may experience increased operational problems such as automatic shutdown of the combustion appliance if the ash builds up into the combustion chamber. Conversely, silica is generally not a problematic element for commercial combustion boilers.

Paulrud et al., (2001), working with reed canary grass, found that the relative content of K and Ca in the ash was more important for agglomeration and clinker formation than the silica content. High-ash agro-pellets (approximately 5% ash) with low to moderate levels of aerosol forming compounds are readily burned in most coal boiler technologies and greenhouse producers in Canada are now installing multifuel boilers capable of burning both coal and agro-pellets.

A comprehensive strategy will be required to reduce the silica content of grasses to make them more convenient for combustion applications and to improve their energy content. The understanding of silica uptake into the plant is improving amongst agronomists and plant breeders. The main cultural factors which appear to have potential to reduce the silica content are: soil type, production region, photosynthetic cycle of the biomass crop and the choice of grass species and variety. The main breeding strategies to reduce silica content include increasing the stem to leaf ratio of the species and reducing silica transport into the plant. As well, fractionation of plant components can help create lower silica containing feedstocks.

The translocation and deposition of silica in plants is heavily influenced by the soluble levels of silica in the soil, present as monosilicic acid or $Si(OH)_4$ (Jones and Handreck, 1967). Clay soils have higher monosilicic acid levels than sandy soils, and therefore produce feedstocks with higher silica levels.

A Scandinavian study found silica levels in reed canarygrass to be highly influenced by soil type; reed canarygrass had silica levels of 1.3%, 1.9% and 4.9% on sandy, organic, and clay soils, respectively (Pahkala et al., 1996).

In Denmark, high silica contents in wheat straw were strongly correlated with clay contents of soils (Sander, 1997). A main difference in silica content between perennial grass species can also be the photosynthetic mechanism of the grass and the amount of water being transpired by the plant.

Warm season (C4) grasses on average, use half as much water as C3 grasses per tonne of biomass produced (Black, 1971). The decreased water usage reduces the uptake of silicic acid and decreases the ash content of the plant. Within warm season grasses, water use per tonne of biomass produced is highest in regions which have a low rainfall to evaporation ratio, and where biomass crops are grown on marginal soils (Samson et al., 1993; Samson and Chen, 1995).

A combination of these conditions may explain some of the higher values obtained by a survey from the United States reporting switchgrass ash contents of 2.8–7.6% (McLaughlin et al., 1996). Regions with a rainfall to evaporation ratio greater than 100% would be expected to have substantially lower ash contents than short grass prairie regions where the rainfall to evaporation ratio is 60%. This is illustrated in analysis from Quebec and Western Europe where silica levels of lower than 3% are commonly obtained in overwintered materials. Plant species have widely differing levels of silica. By comparing the speed of silica uptake with that of water uptake, three modes of silica uptake have been suggested by Takahashi et al., (1990).

These modes are active (higher than water uptake), passive (similar with water uptake) and rejective (slower than water uptake). However, Van Der Vorm (1980), found no evidence of passive uptake. A gradual transition was found between metabolic absorption to metabolic exclusion which depended on the silica concentration. In all species examined, including 3 monocots (rice, sugar cane and corn), there was preferential absorption at low concentrations and exclusion at high concentrations (Van Der Vorm, 1980).

As silica uptake by rice is significantly higher than other agronomic species, considerable efforts and achievements have been made in understanding and characterizing the process. This now has included molecular mapping studies of the silica transport mechanism (Ma et al., 2004). It may be possible that some reductions in the silica content of warm season grasses could be made in warm season grass breeding programs by reducing silica transport into the plant. It should however be noted that sugar cane and rice plant breeders are currently trying to increase the content of silica in these species because silica plays an important role in reducing plant stresses, increasing resistance to diseases, pests, and lodging, and decreasing transpiration (Ma, 2003).

Silica is mainly deposited in the leaves, leaf sheaths and inflorescences of plants (Lanning and Eleuterius, 1989). Lanning and Eleuterius (1987) working in Kansas prairie stands found switchgrass silica contents to be lowest in stems and higher in leaf sheaths, inflorescences and leaf blades. Silica levels are suggested to have evolved to be high in inflorescence structures to prevent the grazing of seed heads. Due to the low stem silica content, the overall silica concentration of grasses decrease as the stem content increases. Pahkala et al., (1996) examined 9 different varieties of reed canarygrass and found varieties to range from 2.3% to 3.2% silica content, with the lower silica containing varieties having a higher biomass stem fraction. Thus, selection for increased stem content is desirable for improving biomass quality for combustion purposes. The impact of ash content on the energy content of the feedstock is evident as the leaves also contained approximately 6% less energy than stems. Stems contained on average 19.55 GJ/ODT which is 98% of the average energy content of high quality wood pellets of 20 GJ/ODT (Obernberger and Thek, 2004).

The differences in silica content between the various components of grasses has been known for more than 20 years. It also appears there are substantial inherent differences between the silica contents of warm season grass species. Two of the 3 main tallgrass prairie species in North America are big bluestem and switchgrass.

The overall silica content of big bluestem may be amongst the lowest of the native North American grasses. In studies of plants harvested from a native prairie, relatively low silica contents of 0.29, 1.69, 2.08, and 2.89% were reported for the stems, leaf sheaths, inflorescences and leaves, respectively.

In contrast, switchgrass averaged 1.03, 3.89, 3.41 and 5.04% for stems, leaf sheaths, inflorescences and leaves, respectively (Lanning and Eleuterius, 1987). As switchgrass is known to grow in wetter zones in the prairies, the higher levels of silica found may be a result of where the plants were collected within the prairie remnant.

Big bluestem is known to have the additional advantage of having a high percentage of its dry matter in the stem fraction and a smaller inflorescence than native ecovars of switchgrass. Typically, the stem fraction of mature native big bluestem ecovars (e.g. cultivars not selected for forage quality) is approximately 60% of the above ground biomass, while in upland switchgrass ecovars the stem typically comprises 45–50% of the biomass in mature plants (Boe et al., 2000; Samson et al., 1999a). Further analysis of species and components of grasses as well as cultivars of grasses is required to more effectively understand how to reduce silica levels.

In the search for low silica herbaceous feedstocks for the pulp and paper industry, there has been considerable research and commercial development in Scandinavia on fractionation technologies to separate the low silica containing stems from the other plant components (Pahkala and Pihala, 2000; Finell et al., 2002; Finell, 2003). Several approaches to dry fractionation have been developed and integrated into commercial straw pulping facilities in Denmark (Finell et al., 2002). The basic process of disc mill fractionation developed by UMS A/S in Denmark is overviewed by Finell (2003) and includes keys steps of bale shredding with a debaler, hammer milling, disc milling, pre-separation (separating leaf meal and internode chips) and then a final sifting to further refine the accepted fraction of internode chips for pulping. In the case of reed canary grass, typically 40–60% of the plant could be recovered for pulping applications with the residual material used as a commercial pellet fuel (Finell, 2003).

This technology can also be applied to the fractionation of warm season grasses to developing fuels for use in the residential and commercial pellet markets. Fractionation of stems from species such as big bluestem would produce pelletized fuels in the range of 1% ash if the feedstock was grown on sandy soils in regions with a favourable rainfall to evaporation ratio. The higher-ash leaf, leaf sheath and inforescence material could then be used as a high-ash commercial pellet fuel for larger-scale thermal applications.

Outlook

This review supports other recent studies that have found energy crop development for thermal energy applications holds significant potential for industrialized nations as a means to create energy security and clean energy through GHG mitigation. From an energy security standpoint, it appears that the conversion of whole plant biomass from annual C4 grasses into biogas or bioheat represent the most promising energy production technologies available. With current understanding of the GHG mitigation issue, direct combustion applications of perennial grasses to displace coal, natural gas and heating is the leading strategy to use farmland to mitigate greenhouse gases. The large N2O emissions associated with the cultivation of corn in humid temperate climates impairs the effectiveness of corn as a feedstock to produce low GHG loading gaseous and liquid biofuels. In this respect, more research on N-efficient annual crops and higher digestibility perennial biogas species could help strengthen the GHG mitigation potential of biogas from energy crops in the future. In the case of bioheat from grasses, the research challenges ahead include the improvement of biomass quality to develop pellet fuels with low contents of silica and aerosol-loading elements.

Some of the largest hurdles to overcome in the emergence of second generation bioenergy technologies are not technological issues, but rather policy barriers. Governments have a major influence on which crops and technologies are scaled up for commercialization through the use of incentives or subsidy programs. It would be highly recommended to encourage policies to avoid picking technology winners in the development of energy security and greenhouse gas mitigation technologies from RET's. Rather, governments should encourage results-based management approaches to address policy issues and examine means to create parity in incentives in the green energy marketplace. This could include the creation of carbon taxes, green carbon incentives, CO_2 trading systems or incentives per GJ of energy produced. Both progressive policy and technology development need to be developed together for renewable energy to work for environmental protection and energy security in industrialized nations.

Bibliography

ARMSTRONG A.P., BARO J., DARTOY J., GROVES A.P., NIKKONEN J., RICKEARD D.J., Energy and green house gas balance of biofuels for Europe, Brussels, CONCAWE, 2002.

BARRACLOUGH S.L., GHIMIRE, K.B.: Agricultural expansion and tropical deforestation. Poverty, International Trade and Land Use, London, Earthscan, 2000.

COURTNEY R.S.: The future of biofuels: a blend of hope and concern. The Centre for Science and Public Policy, Washington D.C., 2006.

KAVALOV B.: Biofuel Potentials in the EU. Institute for Prospective Technological Studies, Joint Research Center, 2004.

KOJIMA M., MITCHELL D., WARD W.: Considering trade policies for liquid biofuels. Washington, World Bank, 2007.

RYAN L., CONVERY F., FERREIRA S.: Stimulating the use of biofuels in the European Union: implications for climate change policies. Energy Policy, 2006.

RUSSI D.: Social Multi-criteria Evaluation and Renewable Energy policies: Two case studies. PHD thesis, Programme in Environmental Sciences, Autonomous University of Barcelona, 2007.

SCHNEPF R.: European Union Biofuels Policy and Agriculture: An Overview, CRS Report for Congress. Congressional Research Service, 2006.

UN-ENERGY: Sustainable Bioenergy: a framework for decision makers, New York, UN-Energy, 2007.

VON BRAUN J., PACHAURI R. K.: The promises and challenges of biofuels for the poor in Developing Countries, Washington DC: IFPRI, 2006.

Index

♦♦♦